Computer-Aided Modelling of Stormwater Systems

Computer-Aided Modelling of Stormwater Systems deals with the construction of probabilistic models of rainfall and their use in the sizing of retention tanks. This book discusses methods of sensitivity and uncertainty analysis of hydrodynamic models of urban catchments, as well as proprietary calculation algorithms for the sizing of retention tanks, considering the uncertainty of calibrated parameters and the variability of outflow hydrographs from the catchment and shape parameters. This is important from the point of view of contemporary hydrology of urban catchments, because it is the reliability of the identified parameters and their appropriate selection that determines the developed hydrodynamic model of the catchment, so that it finds a practical application in sewage network management. Also included are innovative graphical and analytical methods of tank sizing, whereby the shapes of hydrographs of the outflow from the catchment area are considered, which are omitted in most of the methods used. Hybrid models, which are a combination of graphical and analytical methods with probabilistic models of rainfall, are also presented, which make it possible to shorten the calculation time. Finally, the ability to implement computational algorithms in the field of sensitivity and uncertainty analysis can also be used when analysing catchments with implemented green infrastructure facilities.

Computer-Aided Modelling of Stormwater Systems is aimed at doctoral students, research workers, and engineers as an introduction to modern methods of modelling stormwater and combined sewer networks, together with the objects located on them, such as retention tanks.

Computer-Aided Modelling of Stormwater Systems

Advanced Tools Supporting Sustainable Development of Urbanised Catchments

Bartosz Szeląg, Adam Kiczko,
Grzegorz Łagód, Marek Sokáč,
Yvetta Velísková, Cecilia Bruni,
and Francesco Fatone

Routledge
Taylor & Francis Group
LONDON AND NEW YORK

First published 2025
by Routledge
4 Park Square, Milton Park, Abingdon, Oxon, OX14 4RN

and by Routledge
605 Third Avenue, New York, NY 10158

Routledge is an imprint of the Taylor & Francis Group, an informa business

Doskonała Nauka

Ministry of Education and Science
Republic of Poland

The project is co-financed from state budget funds allocated by the Minister of Education and Science of Poland under the Excellent Science II Program.

Funding body: Minister of Education and Science, Lublin University of Technology

British Library Cataloguing-in-Publication Data
A catalogue record for this book is available from the British Library

ISBN: 9781032898537 (hbk)
ISBN: 9781032901756 (pbk)
ISBN: 9781003546436 (ebk)

DOI: 10.1201/9781003546436

Typeset in Times New Roman
by codeMantra

Contents

Author biographies

Bartosz Szeląg is the author of the doctoral dissertation "Influence of the shape of the stormwater inflow hydrograph on the capacity and selection of downstream facilities". He is a specialist in the field of urban catchment modelling using hydrodynamic and statistical models. He has authored works on the development and calibration of hydrodynamic models for quantitative and qualitative forecasting using probabilistic methods (GLUE). He has also developed numerous probabilistic models for sizing retention reservoirs and overflows, taking into account stochastic rainfall characteristics and uncertainty parameters of hydrodynamic models.

His current research directions focus on the implementation of statistical methods for the evaluation of stormwater network operation in the aspect of its potential modernisation and the development of methods for dimensioning facilities in stormwater systems, taking into account the dynamics of rainfall changes over many years, rainfall origin and changes in area imperviousness. He also works with applications of machine learning methods to model the operation of sewage networks and the quality of wastewater, taking into account the uncertainty in the context of urban catchment management and the optimal selection of green infrastructure facilities. He also conducts research in the field of modelling and control of technological processes in wastewater treatment plants. He has authored/co-authored more than 82 publications indexed in Web of Science, 90 publications indexed in Scopus, and 2 monographs.

Adam Kiczko is the author of the doctoral dissertation "Controlling the Siemianówka Dam Reservoir taking into account the environmental needs of the Narew Valley". He has professional experience in developing hydro-engineering projects. He is a specialist in the field of uncertainty analysis in mechanistic and statistical models, as well as control and stochastic optimisation under uncertainty. He is also a programming specialist (MATLAB, C++, Octave, Python). He co-authored works on calibration and sensitivity analysis of hydrodynamic models for urban catchments, river systems, open channels, and numerous studies on water management. His current research is focused on creating probabilistic models for reservoir design, taking into account the uncertainty in hydrodynamic models, as well as CFD and numerical modelling of transient conditions. He has authored/co-authored more than 53 publications indexed in Web of Science, 43 publications indexed in Scopus, and 1 monograph.

Grzegorz Łagód is the author of the doctoral dissertation "Modelling of sewage biodegradation in gravity sewer system". He has a scientific background related to modelling changes in wastewater quality in sanitary, combined sewer and stormwater systems, wastewater treatment plants, and receiving waters, as well as the application of GIS for the development and calibration of stormwater system models. He is a co-author of works on calibration and sensitivity analysis of hydrodynamic models of urban catchments, the influence of processes occurring in the sewage system on the operation of wastewater treatment plants, and the influence of wastewater treatment plants and stormwater drainage on receiving waters. He specialises in the application of bioindication and electronic sensing methods in water and wastewater quality assessment, as well as analysing the performance of wastewater treatment plant bioreactors. His current research directions focus on the development of methods useful for the evaluation of sewer network performance in the aspect of their potential modernisation due to changes in catchment development, as well as the quantity and quality of transported wastewater and sediments. He has authored/co-authored more than 166 publications indexed in Web of Science, 168 publications indexed in Scopus, and 11 monographs.

Marek Sokáč is a graduate from Slovak University of Technology, Bratislava, with a specialisation in Hydraulic Engineering and Water Resources Management. In 2009, he received the associate professor degree for his thesis titled "Modelling the impacts of discontinuous pollution sources on surface water". He participated in several study visits and postgraduate courses in Denmark, the Netherlands, Italy, Belgium, and Canada. He has been a principal investigator and co-worker in more than 15 national and international projects and a local team deputy leader on two EU-funded projects (7FP and Horizon 2020). He has participated in more than 50 technical design or expertise projects. Between 2003 and 2008, he was a project manager of the infrastructure project "Environmental improvement of the Liptov Region" FM 2002 SK 16 P PE 009 (10.52 mil. €, construction of sewer systems, total 72 km). His research expertise revolves around sewer systems, urban drainage, urban hydrology, water quality, pollution spreading in water bodies, and numerical simulation. He is currently engaged in the localisation of pollution sources based on monitoring data (inverse task). He has authored/co-authored more than 200 publications, including 3 monographs, 2 chapters in monographs, 22 publications indexed in Web of Science, and 33 publications indexed in Scopus.

Yvetta Velísková is a graduate from Slovak University of Technology, Bratislava, with a specialisation in Hydraulic Engineering and Water Resources Management. In 1995, she received a PhD degree with the thesis "Judgement of surface water dispersion characteristics with 2-D idealization". Her research interests include surface water and groundwater flow, their mutual interaction, pollution spreading in water bodies, numerical simulation of those phenomena, hydrodynamics of flow with a free water level (as in sewer systems), hydrology, and water quality problems. She was a principal investigator and co-worker in more than 20 national and international projects and expertise assignments, as well as an institutional team leader in two EU-funded projects within the 7FP and Horizon2020 programmes. She has been involved with more than 200 scientific publications, including 2 monographs, 5 chapters in monographs, 45 publications indexed in WOS, and 67 publications indexed in Scopus.

Cecilia Bruni conducts studies in the Environmental−Chemical Engineering sector at Marche Polytechnic University, Italy. Her research revolves around the treatment and valorisation of wastewater, organic waste, and secondary sewer flows, for the purpose of maximising recovery processes and minimising environmental impacts. She also works in the field of sustainable development of urban catchments, aimed at improving the operation of sewage networks and reducing the amount of pollutants reaching the receiver. She has authored/co-authored 6 publications indexed in Web of Science, 6 publications indexed in Scopus, and a monograph chapter.

Francesco Fatone is a specialist in the field of water and wastewater management. He has co-authored numerous solutions in the field of sustainable development of urban catchments, aimed at improving the operation of sewage networks, reducing the amount of pollutants reaching the receiver, and minimising greenhouse gas emissions to the atmosphere. He is a specialist in the field of advanced technological processes implemented in wastewater treatment plants (removal of biogens, microplastics, xenobiotics, industrial substances) and the optimisation of these processes based on computational systems in the field of the facility's life cycle, taking into account technical, environmental, and social aspects. He has authored/co-authored more than 93 publications indexed in Web of Science, 148 publications indexed in Scopus, and 10 chapters in monographs.

Introduction

The progressive urbanisation and climate changes lead to an increase in the value of maximum flows, as well as the volume of wastewater outflow from combined sewer systems or stormwater sewer systems. This results in an increase in the number of stormwater discharges (Abdellatif et al. 2015; Kleidorfer et al. 2009), an increase in the frequency of stormwater flooding, and deterioration of water quality in receiving waters (Kirshen et al. 2015; Tavakol-Davani et al. 2016; Szeląg et al. 2020). The large scale of the problem is confirmed by the large number of national and foreign studies in this field (Arnbjerg-Nielsen et al. 2013; Arnell 2011; Bates et al. 2008). To limit the problems with the operation of sewer systems, it is necessary to take adequate measures to delay the stormwater flow (retardation) and reduce its volume and maximum flow rates (reduction). As shown by literature review (Todeschini 2016; Wu et al. 2013), in order to improve the operation of stormwater and combined sewer networks, green infrastructure systems are designed within partial catchments intended for the reduction of stormwater volume, which is currently the subject of many studies (Baek et al. 2015; Eshtawi et al. 2016). An alternative solution is offered by retention tanks, as demonstrated by various investigations (Słyś 2010; Mrowiec 2009). However, these solutions must be properly selected, and their structures and sizes must be optimised to fulfil their tasks and yield satisfactory results (Dziopak & Słyś 2006; Stirrup & Marchandt 2002). Therefore, it is necessary to simulate their operation for a given stormwater load (Balistrocchi et al. 2009; Kisiel et al. 2008; Becker et al. 2001; Andrés-Doménech et al. 2010; Calabro 2004). A key element constituting the basis of the simulation of the operation of a reservoir, sewer network, or its overflow volume is the transformation of rainfall into the so-called outflow hydrograph. Hence, better representation of stormwater flowing through the sewer system will yield a more accurate and correct assessment of the function of its objects. On the basis of the literature data (Baker 1979; Basha 1995; Graber 2009; Hong 2008), it can be concluded that simplified methods can be used for modelling/designing stormwater sewer systems in the case of small urban catchments. In these methods, the modelling of wastewater outflows and flows through the stormwater sewer system is simplified, which has a significant impact on calculation results. The simplified approaches are based on rational methods, e.g., constant rainfall intensity, threshold rainfall intensity, and delay factor methods, and their modifications (Akan & Houghtalen 2003). However, the results of calculations conducted with the rational method are merely estimates (Dayarante & Perera 2004; Dąbrowski 2004; Kotowski 2011). In engineering practice, hydrodynamic models are more frequently used at the stage of generation of models of stormwater sewer networks to improve the credibility of calculation results and assess the operation of the systems (Szeląg et al. 2022b; Rossman 2010; Venutelli 2011; Zoppou 2001; Rossman 2020). These models facilitate simultaneous reflection of non-homogeneous rainfall distributions within the catchment, unsteady flow conditions in the sewer network, potential stormwater outflows onto the land surface, types of land use in the catchment area, and the effect of stormwater accumulation within the network on outflow variability. Additionally, an important advantage of the models is the possibility to simulate the impact of objects designed in the network (retention tanks, stormwater flooding, flow regulators, etc.) on its function. Hence, given their advantages, the models are increasingly being used to design and model the operation of sewer networks, as confirmed by the large number of studies in this field (Fu et al. 2011; Jato-Espino et al. 2018; Martínez-Solano et al. 2016; Djordjević et al. 2005; Bellos & Tsakiris 2016). However, many detailed data have to be collected to model the catchment area. These include the type of land use, terrain features, and sewer system parameters (diameters, conduit slopes, and manhole depth). These data are indispensable for the use of the hydrodynamic model. Nevertheless, the model must be calibrated to produce calculation results that will be consistent with the results of observations of the analysed sewer system (Lee et al. 2018; Li et al. 2014a,b; Martínez-Solano et al. 2016; Niazi et al. 2017). This requires data from high-resolution measurements of both rainfall and flows, with a time step not exceeding 5 min for rainfall and not longer than 1 min for flows (Bruni et al. 2015; Price 2011; Muleta et al. 2013; Skotnicki & Sowiński 2015). Nevertheless, the available results show that there may be problems with the calibration of the model. Firstly, the simulation outcomes may not be consistent with the measurement results (Han & Burian 2009; Zhan & Huang 2004). Secondly, computer simulations (Caradot et al. 2020; Van Bijnen et al. 2012) indicate that calculation results consistent with measurements are obtained in the case of many different combinations of parameters. In practice, this means that the parameters are biased with uncertainty, which has an impact on the calculation of the variability of the stormwater outflow and the sizes of objects designed in stormwater sewer systems. This generates the necessity to develop methods for modelling and dimensioning drainage systems, taking into account the uncertainty of calibrated catchment models, as well as methods in which the uncertainty is minimised.

This book presents several viewpoints on modelled phenomena related to sewer network functioning. It reports on the basics of sewer network design and objects located therein. It also presents a modern approach to the use and calibration of rainfall-runoff models based on mechanistic models, which is currently the subject of numerous studies, scientific dissertations, and research projects. Bearing in mind the usefulness of the presented tools, examples of model use are shown, which have not been described in academic textbooks on modelling urban catchments to date. This book presents the current trends in the study of the influence and interaction between parameters calibrated in mechanistic models on simulation results, using examples of urban catchments in Poland, Slovakia, Croatia, and Italy. This approach is presented through uncertainty analysis based on the GLUE method, which is the subject of numerous analyses of process modelling in environmental engineering, process chemistry, biotechnology, etc. As an illustration of the issues presented above, original solutions and aspects of the application of uncertainty analysis results in dimensioning retention reservoirs and determining the duration of design rainfall were discussed. The latest calculation methods for quick estimation of tank capacity and potential correction with machine learning methods are presented, which have not been described in detail so far. The theoretical basis of the method of analysis and design of stormwater basins from an overflow aspect was also presented, taking into account the variability of the catchment runoff hydrogram and the stochastic nature of rainfall, which have not been comprehensively studied to date and can be applied to Central Europe (Poland, Slovakia, Croatia, Czech Republic, Germany, Austria, etc.), Western Europe (Italy, France, Spain, Denmark, etc.), America, Africa, and Asia. The approaches and issues discussed in this book constitute the basis and introduction to advanced computational methods that may be useful for engineers, doctoral students, and researchers.

Chapter 1

Subject of urban hydrology

1.1 Definition of urban hydrology

Urbanisation in some parts of the Earth has a history of more than 6,000 years, but its urban development did not change dramatically until the 19th century as a result of the Industrial Revolution. In the second half of the 19th century, industry and public transport started to develop, which led to further lifestyle changes. The streets were gradually strengthened, larger buildings were built, and large cities began to form. However, the emergence of new cities and the rapid industrialisation, combined with the disorganised sanitary conditions of the cities, caused various problems, such as plague, typhus epidemics, etc.

Urbanisation can be defined as the formation and development of the urban way of life and/or the increase in the number of people living the urban lifestyle. This also comprises the process of centralising life into the creation and growth of the population living in cities, and the penetration of urban elements into the whole projection settlement.

From 1800 to 1950, the number of people living in cities around the world increased from less than 40 million to more than 700 million. In recent decades, further rapid population growth has been witnessed, especially in developing countries in South America, Asia, and Africa. In particular, there has been a significant increase in population in large cities.

The overall percentage of the inhabitants living in urban areas is shown in Figure 1.1.

UNESCO (1979) defines hydrology as a physical science that deals with all the waters of the Earth, their occurrence, cycle and distribution, chemical and physical characteristics, and their interaction with the environment. This definition has at least two aspects: first of all, it highlights the interdisciplinary nature of hydrology, which deals with the physical, chemical, and biological characteristics of water, and also the temporal and spatial distribution and movement of water in all its forms. The second aspect can be represented by the "classical" hydrological cycle of water on Earth. Figure 1.2 shows the hydrological cycle in systemic terms, indicating the interactions between the various factors of the water cycle on Earth.

The hydrology of urbanised areas is a relatively young scientific discipline. It dates back to the late 1950s and early 1960s and is linked to mass production and transportation development in the United States. Roads began to be strengthened, large residential areas were created around the city centres, and cities with several central parts were created. The consequence of this expansive urbanisation was large cities with a large share of impervious areas established, and problems with rainwater drainage. Therefore, in the early 1960s, experts began to address

DOI: 10.1201/9781003546436-1

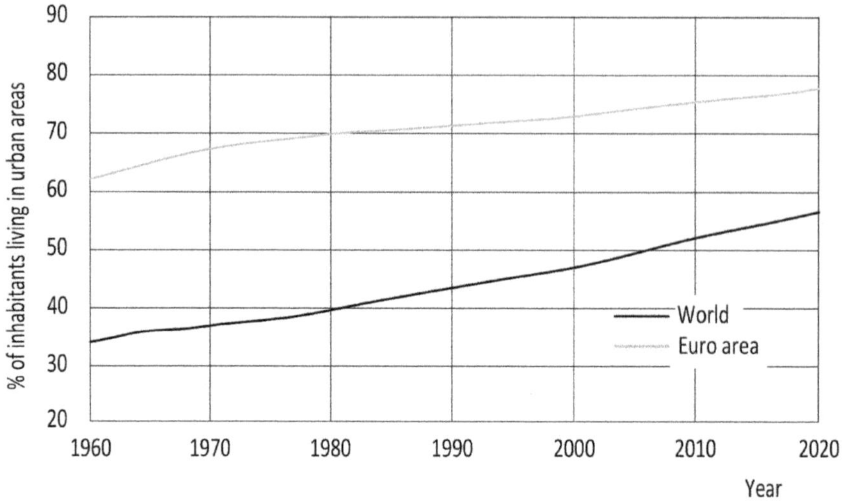

Figure 1.1 Percentage of the inhabitants living in urban areas. (Data source: https://ourworldindata.org.)

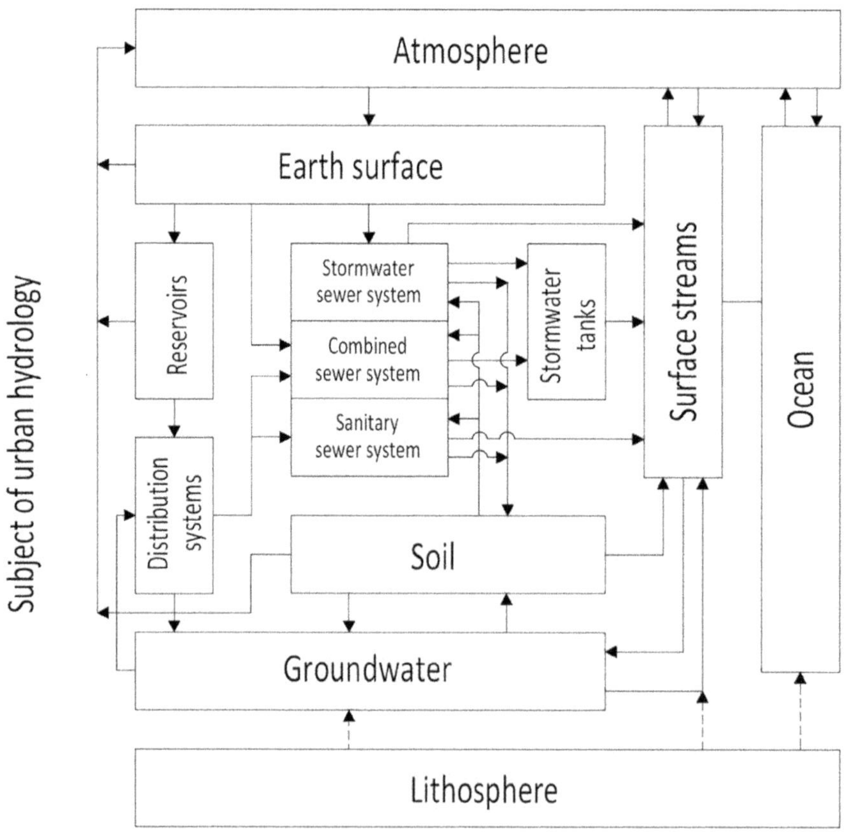

Figure 1.2 Water cycle in urban areas. (Based on Hall 1984.)

the quantitative problems associated with rain runoff. The first mathematical dependences of model rain on the dimensioning of rainwater drainage networks were formulated. Later, due to the sharp deterioration of the water quality of the recipients, attention was also focused on solving the capture and disposal of stormwater, runoff retention in the river basin and sewers, stormwater treatment, etc. Today, rain runoff from urbanised areas is assessed using an integrated approach to the sewer network–recipient–WWTP.

The hydrology of urbanised areas, as follows from its definition, differs from the hydrology of natural river basins in the spatial and temporal scales of the studied phenomena. The size of the river basin and typical flow times are significantly smaller in the hydrology of urbanised areas than in natural river basins (Figure 1.3).

1.2 Changes in the natural urban water cycle due to urbanisation

Urban areas disrupt the natural hydrological regime by building engineering works. Cities are protected against waters outside the urban areas by a network of catch trenches; surface water from impervious areas is drained by a sewer network; open or underground stormwater

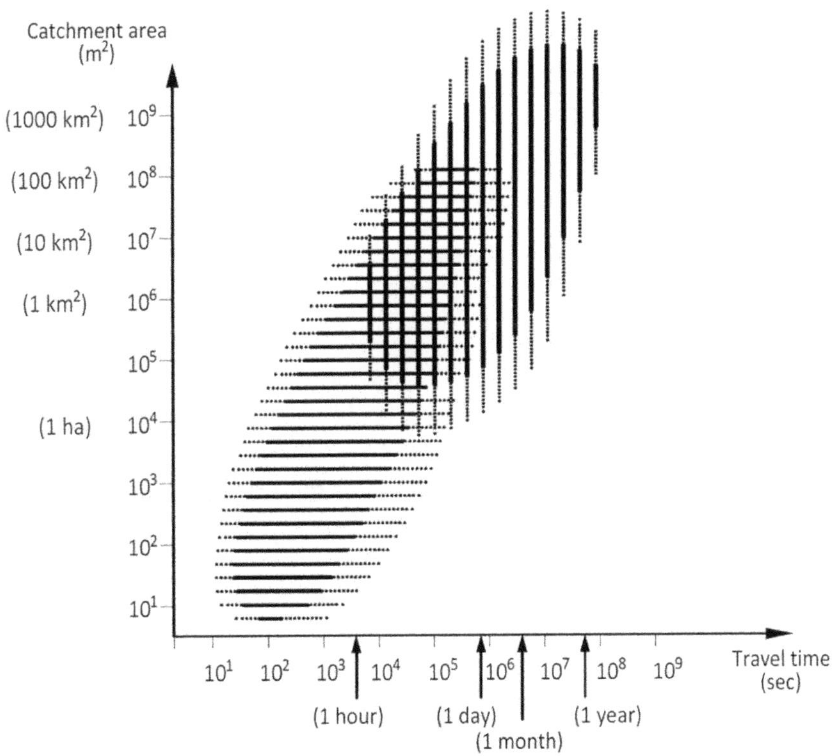

Figure 1.3 Different areas of hydrological processes in natural and urban areas.

reservoirs are built to temporarily accumulate runoff; natural replenishment of groundwater reserves is reduced; groundwater is used for drinking purposes; wastewater is discharged into watercourses, and eventually to the seas and oceans.

There are a number of elements in the hydrology of urbanised areas that change the nature of runoff from river basins. The most important differences between runoff from natural and urbanised river basins are as follows:

- Unlike natural river basins, urbanised river basins are subject to constant change, which also causes changes in the runoff regime of stormwater. Due to these constant changes, statistical methods can only be used to a limited extent when evaluating long-term sewerage network records or trends.
- In urbanised river basins, there is a substantial increase in impervious surfaces. The more impervious areas there are in the basin, the greater the part of the rainwater runoff that will enter the sewer network (Figure 1.4). This leads to an acceleration of the runoff process.
- Higher outflow velocities lead to faster runoff concentration and faster occurrence of maximum flows; on the contrary, the total runoff duration is shorter (Figure 1.5).
- A larger volume of water flows through the sewer network in a shorter time, which leads to an increase in maximum flows and increased sediment transport in the sewer network, as well as in the recipient. The consequence of increased flows in the recipient is bottom and bank erosion, and subsequent deposition of resuspended sediments, as well as morphological changes in the recipient. Some of these factors are short-term (e.g. increased flows), while others are long-term (e.g. morphological changes).
- Total runoff volumes are also greater compared to natural river basins. This increase can be up to 50%.

Figure 1.4 Water balance of unpaved and paved areas. (Based on Sokáč 2019.)

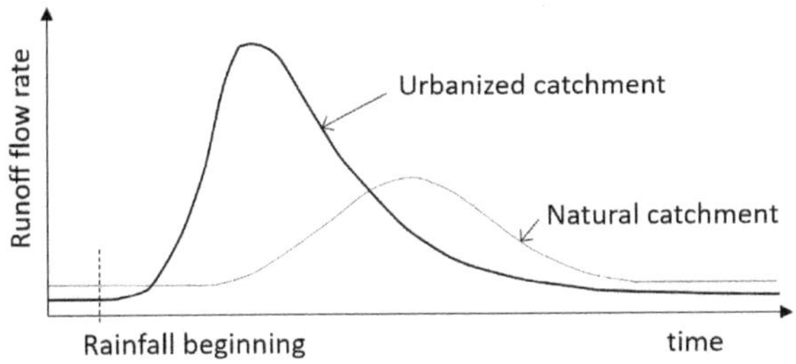

Figure 1.5 Hydrological runoff from natural and urbanised river basins.

- Figure 1.5 demonstrates that, in urbanised river basins, there is a smaller base runoff in surface flows due to the rapid runoff of stormwater from the river basin. It is this phenomenon that has the most adverse environmental effects.
- In addition to the short-term effects mentioned, urbanisation also has long-term effects on the hydrological regime of the river basin, which are manifested, e.g., in draining large flows with a higher periodicity of occurrence, which reduces the possibility of their use, e.g., for recreational or sports purposes.
- Another accompanying phenomenon of urbanisation is the temperature increase in the recipients due to the heating of rainwater on paved areas. Even a 10% increase in impervious areas can lead to an increase in the temperature of watercourses due to the inflow of water from these areas, while the outflow from overheated impervious areas can be up to 10 °C warmer than the outflow from natural river basins.
- Urbanisation causes increased soil erosion. In principle, this corresponds to surface erosion, which acts on most impervious surfaces and is usually accompanied by concentrated erosion in canals and potholes. Increased sediment transport from the river basin surface creates problems in sewer networks, especially in flat areas with pipes laid at minimal slopes.
- The increase in impervious areas leads to a reduction in the natural infiltration of stormwater into the deeper underground layers. This results in a reduction in the groundwater level, especially with the intensive use of groundwater to supply the population with drinking water. Soil drying can also have serious economic consequences, especially in areas with clay soils, where clay drying can lead to soil settling. Open stormwater tanks built on a divided rainwater drainage system, with subsequent infiltration of accumulated water, can partially assist in groundwater recharge.

The impact of urbanisation on the hydrological cycle extends beyond the boundaries of urbanised settlements and has at least a regional character, affecting the fauna and flora in the adjacent river basin. The hydrological cycle of urbanised areas is of a complex nature, where not only the elements of runoff accumulation have a diverse character, but also their mutual connections are more complex than in the natural hydrological cycle. Another, perhaps the most important, factor in terms of environmental protection and ecological balance, which is the pollution of stormwater and recipients, enters this complex cycle.

Stormwater pollution is caused by two factors: primary air pollution due to the leaching of pollutants from the atmosphere, and secondary pollution during runoff on the earth's surface, respectively. Polluted air denotes the state of the atmosphere when there are sufficient substances present in the air for a shorter or longer period of time that adversely affect the environment. In this case, an air pollutant is defined as an ingredient which, either directly or after changes in the atmosphere, damages or endangers living organisms and adversely affects the environment (Tölgyessy et al. 1989).

Thus, atmospheric pollution expresses the sum of solid, liquid, and gaseous pollution in the air that is washed away by rain or snow. The air is polluted either by the release of various substances into the atmosphere or by events taking place directly in the air (e.g. chemical reactions). Pollutants are divided into primary and secondary, depending on the place of origin. Unlike primary pollutants, which enter the air from natural sources or human activity, secondary pollutants are formed directly in the air due to atmospheric reactions. These reactions primarily include photochemical reactions, oxidation and reduction processes, polymerisation, and catalytic processes. As a result of these reactions, secondary pollutants are sometimes produced with a higher harmful effect than primary emissions. Rainfall, in the form of raindrops or snowflakes, washes away much of the solid, liquid, and gaseous atmospheric impurities from the air by dissolution, adsorption, and absorption processes. Some of the dusty impurities, which act as condensation cores at sufficient humidity, affect not only the composition of the rainfall but also the rainfall formation.

Of the primary gaseous emissions, carbon, sulphur, and nitrogen oxides should be mentioned. The actual amount of gaseous emissions in the atmosphere depends on their reactivity. Most attention is paid to methane (CH_4), carbon dioxide (CO_2), and nitrous oxide (N_2O), which are non-reactive and insoluble gases with a long life, expressed in years. In contrast, other nitrogen oxides (NO_x), ammonia (NH_3), and sulphur dioxide (SO_2) are reactive and more soluble, so they can be washed out of the atmosphere by rainfall, which means their lifespan is only a few days.

Of the secondary gaseous pollutants, tropospheric ozone (O_3) deserves attention, as it is formed by chemical reactions and produces photochemical smog. Sulphuric acid (H_2SO_4) is also an important secondary source of pollution, formed in the atmosphere by the oxidation of discharged sulphur dioxide and, together with nitrogen oxides and ammonia, contributing to the formation of acid rain. Acid rain is one of the most serious consequences of atmospheric pollution. Halogens (chlorine, hydrogen chloride, hydrogen fluoride) are another secondary source of pollution. Chlorine acts as a catalyst in the destruction of stratospheric ozone. Hydroxyl anions (OH^-) are another source of pollution. Although their concentration is negligible, they are important for the oxidation of several primary pollutants and thus help to self-clean the atmosphere due to the flushing of pollutants in the rain. The increase in the concentrations of CO and CH_4 in the atmosphere, with which the OH^- anions react the fastest, therefore has a negative effect on the self-cleaning ability of the atmosphere. As part of various reactions, radioactive substances, which are transferred and constantly changing, are also formed in the air. The systematic study of these substances has intensified in relation to the effects of nuclear tests and nuclear power plants on the composition and changes of the atmosphere and biosphere. The radioactivity of the atmosphere is caused by the presence of radioactive aerosols and gases.

Anthropogenic resources, despite their small share, adversely alter the natural balance created between natural resources and the processes that remove pollutants from the atmosphere. Human activity releases substances into the air that are more aggressive and can seriously endanger many terrestrial biological processes. The great danger of anthropogenic emissions lies in the fact that they are concentrated in significant amounts in the air of industrial areas and urban settlements. Sources of emissions from human activities are divided into stationary (especially the energy industry, industrial technologies, and incinerators, and to a lesser extent households) and non-stationary (transport). The share of transport in total emissions is estimated at around 50%–60%.

Chapter 2

Input data

2.1 Data for developing mechanistic models

The tasks solved within the sewerage of towns and municipalities, including the management of stormwater according to the time horizon of the expected results, are divided into two groups:

- current status,
- the future.

Accurate determination of the initial data for the current and expected prospective state is a basic precondition for the reliable design and assessment of sewer networks. The required accuracy and range of input data are oriented according to the type of calculation task and the degree of detail required.

The following data are required for the hydraulic assessment of sewer networks:

- rainfall data,
- data on the sewerage catchment,
- sewerage network data,
- data about the recipient.

Sewerage and river basin data are a type of data that can be obtained in a relatively simple way (field cataloguing, levelling, tachymetric measurement, download of project data, etc.), while meteorological data (especially rainfall data) require technologically demanding technology and, in most cases, a long observation period (Grimaldi & Serinaldi 2006).

2.2 Rainfall data

Rainfall is the primary input to the models describing surface run-off, stormwater network drainage, and, in the case of hydraulic overload, stormwater flooding. The measurement of rainfall plays an important role, because hydrodynamic models are used to evaluate the performance of the stormwater network, the impact of the stormwater system on the receiver, and to make decisions on potential modernisation (construction of retention reservoirs, construction of GI facilities). Currently, rainfall gauges are the most common devices used to measure rainfall, but their location must meet a number of conditions to ensure high consistency of the rainfall data collected (IMGW-PIB 2015; Michelson et al. 2007). In order for the

DOI: 10.1201/9781003546436-2

obtained measurement results to be reliable, the requirements concerning the device location from the ground level, relief, distance from the heat sources, buildings, roughness of the terrain, distance from water reservoirs, height of the objects protecting the rain gauge exposure, and angle of horizon obscuring by the obstacle in relation to the wind direction and speed measurements, should be met. The performed measurements showed that the wind speed and air movement have a great influence on the reliability of the obtained rainfall measurements (Nešpor & Sevruk 1999; Pollock et al. 2015; Blocken et al. 2013; Hangan 1999). Despite numerous criteria which should be met by rain gauges, attention should also be paid to the device used and its measurement sensitivity. Buoyant flow meters have been replaced by electronic devices, which allow conducting rainfall measurements. At present, trough rain gauges are the most commonly used devices. Owing to proper automation of rainfall data recording, they enable the collection of rainfall data with high resolution. The results of rainfall measurements using weighted, trough, and non-contact rain gauges under comparable conditions were discussed in the papers by Colli et al. (2014). The results of trough rain gauges indicate that they underestimate rainfall at an intensity of 100 mm/h. Higher measurement accuracy with lower operational requirements is provided by weighted rain gauges (Price & Vojinovic 2011).

The location of rain gauges and the resolution of rainfall measurements in urban catchments depend on the purposes for which the data are to be used. Price and Vojinovic (2011), based on a study of a rainfall field in southern France, determined the recommended spatial resolution of rain gauges using the following relationship:

$$\Delta r = 1.5 \cdot \sqrt{\Delta t} \tag{2.1}$$

where: Δt – rainfall measurement resolution (min).

The relationship given above is difficult to generalise to all purposes for which rainfall can be used in urban catchments. The basic engineering tasks (Table 2.1) for which rainfall data are used dictate the resolution and methodology of their measurements:

- Designing sewer networks, which requires the knowledge of maximum flows,
- Checking calculations to verify the capacity of the designed stormwater networks,
- Evaluating the performance of stormwater networks, verifying and calibrating hydrodynamic models,
- Using on-line measured rainfall data to determine the optimal flow control strategy for a stormwater network.

The recommendations that should be met by rainfall data for the respective tasks were established considering the following factors (Cristiano et al. 2017):

- The period of rainfall measurement,
- Distance of the rainfall measuring device from the catchment area,
- The resolution of recorded rainfall,
- Spatial resolution of rainfall,
- Time synchronisation error.

Table 2.1 Rainfall data requirements depending on application in drainage systems (Butler & Davies 2011; Mrowiec 2009, as amended by Szeląg et al. 2020; 2021a)

Issue	*Measurement period, years*	*Location of the rainfall station*	*Measurement resolution*		*Synchronisation error*
			Temporal (min)	*Spatial (km²/station)*	*min*
Sewerage design	> 10				
Stormwater overflow design[c]	> 5[e]	Over 25 km	block	Uniform rainfall in the catchment	< 30
Retention Basin design[a]	> 5		< 15		
Verifying calculations[a]:					
sewers	>20 (10[d])	In vicinity of the catchment	1	Uniform rainfall in the catchment	< 10
Retention basins, overflow structures	>10		5	< 5	< 5
Analysis of the operation of existing stormwater networks[d]	Heavy rainfall caused by convective rainfall	In vicinity of the catchment	2	2	
Calibration, verification[b]	Rainfall events caused by rainfall of identical origin		2	2	0.25
RTC	On-line	In the catchment	2	2	0.25

[a] In the calculations, it is recommended to take into account the interaction between the calibrated parameters of hydrodynamic models (Kiczko et al. 2018; Szeląg et al. 2016; 2019; 2020; 2021a; Andrés-Doménech et al. 2010; Bacchi et al. 2008),

[b] Due to the influence of catchment characteristics as well as rainfall intensity and rainfall distribution on the results of sensitivity analysis, it is advisable to properly select the rainfall – runoff events for model calibration and validation, taking into account the above-mentioned factors (Szeląg et al. 2016; 2020; 2022c; Fatone et al. 2021),

[c] Calculations for 6 years of rainfall data confirmed the usefulness of rainfall series for analysing the performance of a stormwater overflow structure (Szeląg et al. 2016),

[d] Based on rainfall data as short as 10 years, a synthetic rainfall generator can be developed to simulate multi-year rainfall events (Szeląg et al. 2018),

[e] 45-year rainfall series can be used to develop multivariate probability density distribution taking into account local changes in rainfall dynamics and correct the position of the overflow structure edge or catchment characteristics on a long-period basis (Szeląg et al. 2020; WMO-No. 1203 2017).

2.3 Design rainfalls

2.3.1 *Block rainfall*

The rainfall is one of the basic factors exerting a significant impact on the results of simulation of stormwater outflow from the sewer system. The reliability of collected rainfall data is essential for modelling the operation of sewer networks (Mrowiec 2009; Blocken 2013; Dai & Han 2014). The proper location of the pluviograph (Suligowski 2004; WMO 2011; Acquaotta et al. 2016) and the type of the measuring device (Ciach & Krajewski 1999; Ciach 2003; Kusmierek-Tomaszewska 2009; Kotowski et al. 2011; Colli et al. 2014) are important elements in fulfilling this condition. Detailed information on these issues was presented by Licznar (Licznar et al. 2011) and Kotowski (Kotowski 2011). The phenomenon of a single rainfall event is described by its intensity ($I=P_c/t_d$, where: P_c –rainfall depth in mm, t_d is the rainfall duration in hours) in mm/h or the unit rainfall intensity ($q=166.7 \cdot I$) expressed in liters $ha^{-1} \cdot s^{-1}$, rainfall duration (t_d) in minutes or hours, and territorial range. In practice, due to the unavailability of rainfall measurement results, constant rainfall intensity during the rainfall event (rainfall histogram) and within the catchment is often assumed (Figure 2.1). The constant rainfall intensity (so-called block rainfall) is a result of statistical treatment of the rainfall data. The block rain concept was used in the time, when no rainfall-runoff models existed; the concept is nowadays, of course, different. Constant rainfall can be easily transformed into a design rainfall with different shape, for example, triangular, likewise in the stormwater tank. The basic assumption is that the stormwater volume will be the same (for a given frequency and rainfall duration).

Noteworthy is the fact that the rainfall intensity distribution is not included in many calculation methods (Fu & Butler 2014; Grum & Aalderink 1999; Osorio et al. 2009), which is associated with the size of the rainfall dataset. In design practice, to quantify temporal rainfall distribution, regional models are used for determination of the mean intensity or unit intensity of rainfall with an assumed probability of rainfall occurrence. The variability of the unit rainfall

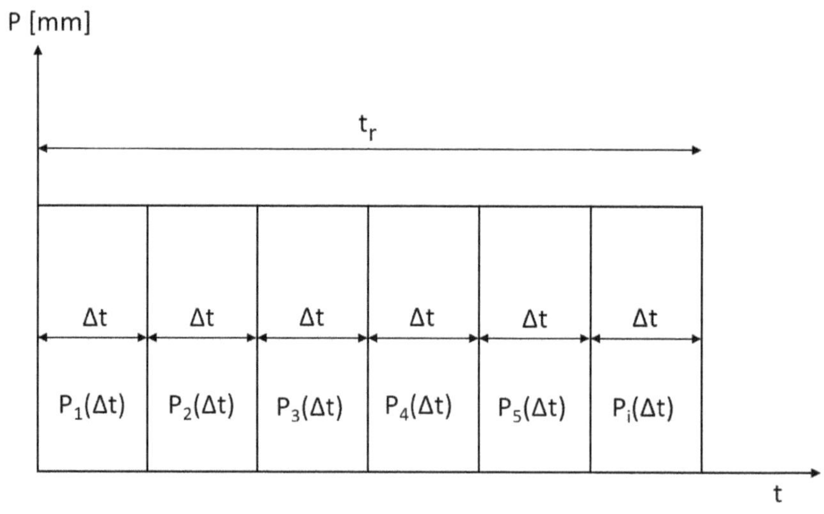

Figure 2.1 Block rainfall distribution.

Notes: Δt – time step, $Pi(\Delta t)$ – rainfall depth in the time step, $\sum Pi(\Delta t)$ – total rainfall depth in the rainfall event.

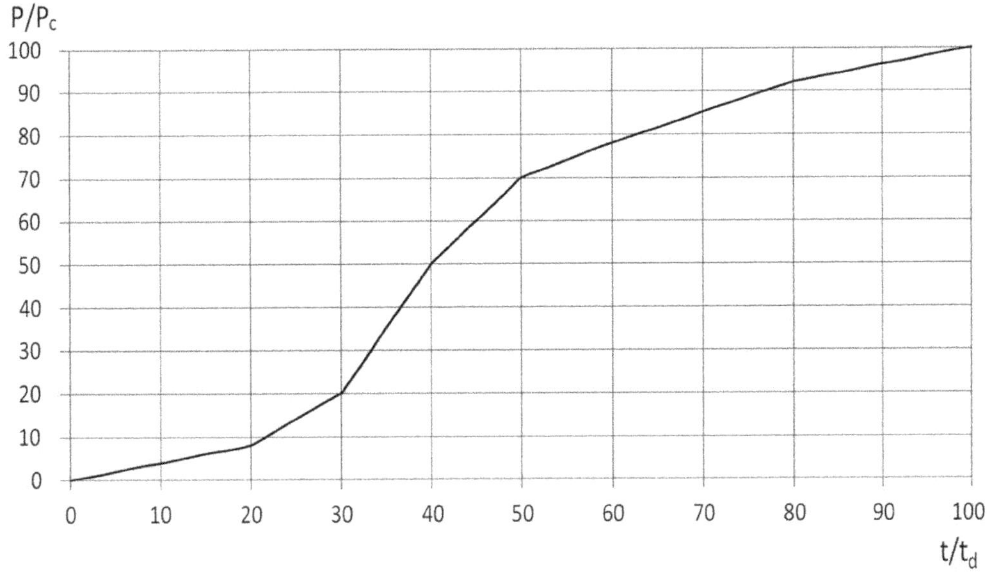

Figure 2.2 Dimensionless cumulative rainfall curve $P/P_c = f(t/t_d)$ describing the Euler type II rainfall (Akan & Houghtalen 2003).

intensity as a function of its duration for the assumed probability of return period is usually illustrated by IDF (Intensity-Depth-Frequency) or DDF (Depth-Duration-Frequency) curves. Rainfall models developed with the use of the curves are the basis for designing a sewer network or its objects. In the case of large sewer systems, Euler type II rainfall models are used for assessment of the frequency of the stormwater flooding phenomenon (Figure 2.2) (Urcikán & Horváth 1984; Knighton & Walter 2016). From the point of view of the design of sewer network objects and the safety of their operation, multiyear measurements of rainfall are recommended, as they will help to determine dimensionless rainfall distributions in a rainfall event with its origin (Hegger et al. 2016; García-Bartual & Andres-Domenech 2017), duration (Wartalska & Kotowski 2020), or regional frequency (Szeląg et al. 2022b; 2021c; 2020).

The duration of the model rainfall (Figure 2.2) used for simulation of the operation of the stormwater sewer system should include all design rainfall durations within the sewer network. The recommendation to include the rainfall duration, which is twice the longest flow time in the sewer network, is prompted by experience gained through application of this method and comparative calculations from simulations of sewer system operation based on multiyear rainfall measurement series. A different approach to determination of rainfall duration is employed for dimensioning retention reservoirs. The design stage involves searching for rainfall duration (so-called design rainfall) for which the determined volume has the maximum value.

Due to the limited access to rainfall data (a multiyear period of 30 years), a number of regional empirical models have been developed to determine the relationship between the maximum amount of rainfall (rainfall intensity), its duration, and probability of occurrence. In Poland, the basic empirical formulas have been proposed by Błaszczyk et al. (1974), as well as Bogdanowicz & Stachy (1998). In both methods, the traits of the "rainfall region" are taken into account indirectly (Błaszczyk) or directly (Bogdanowicz & Stachy). In the former case, rain intensity (q) is calculated from the formula (Błaszczyk et al. 1974):

$$q = \frac{6.631 \cdot H^{0.67} \cdot \sqrt[3]{C}}{t_d^{0.67}} \tag{2.2}$$

which at $H = 585$ mm yields:

$$q = \frac{470 \cdot \sqrt[3]{C}}{t_d^{0.67}} \tag{2.3}$$

where: H – annual rainfall depth (mm), t_d – rainfall duration (min), q – rainfall intensity, $dm^3 ha^{-1} s^{-1}$, C – rainfall frequency assumed for calculations.

As proposed by Bogdanowicz & Stachy (1998), the maximum rainfall depth (P_{max}) with rainfall duration (t_d) and probability of exceedance (p) can be described by the following formula:

$$P_{max}(t, p) = 1.42 \cdot t_d^{0.33} + \alpha(R, t_d) \cdot (-\ln p)^{0.584} \tag{2.4}$$

where: P_{max} – maximum rainfall depth with the assumed rainfall duration (t_d) and probability of exceedance (p), mm, $\alpha(R,t_d)$ – regional parameter calculated depending on the location of the object and rainfall duration.

Research has shown that rainfall depths determined in the BiS model for C = 1 are underestimated *compared with* the calculation results obtained with the use of the Błaszczyk formula (Eq. 2.3). Therefore, this formula has limited use. On the basis of the results of multiyear rainfall measurements for the Suków-Kielce station, Suligowski (2004) proposed the following formula for the determination of the maximum probable rainfall intensity:

$$I_{max} = A_0 \cdot t_d^B \tag{2.5}$$

where: t_d – rainfall duration (min) in the range of 30 ÷ 1,080 min, A_0, B – regression coefficients depending on the probability of exceedance.

A number of local models (Licznar et al. 2018a,b; Suligowski 2004; Bogdanowicz & Stachy 1998) are used for modelling the maximum rainfall depth (rainfall intensity). Since different methodological approaches are employed, which should be standardised together with their results, such an attempt has been made. Currently, a Precipitation Atlas is being compiled in the country as part of the PANDa project (Burszta-Adamiak et al. 2020). The problem addressed by the project seems to be highly complex, as rainfall data may lose continuity in many cases. Importantly, it should be borne in mind at the stage of generation of rainfall models that the measuring devices used by the Institute of Meteorology and Water Management were changed in 2005. This is a very important issue, since the data used for simulation calculations should be homogeneous (Bogdanowicz & Stachý 1998).

2.3.2 *Generators of synthetic rainfall*

Modelling and evaluation of the operation of existing and planned stormwater networks and objects located therein can be carried out with the use of the so-called synthetic rainfall series

(Sanvicente-Sánchez & Solís-Alvarado 2008; Acharya et al. 2017; Wilks & Wilby 1999; Rodriguez-Iturbe et al. 1988). These are rainfall time series yielded by simulation results. It was shown that canonical micro-cascades, machine learning methods, etc. can be used to model rainfall series; those models, usually based on the multivariate probability density distributions, are currently one of the most frequently used methods for simulation of synthetic rainfall series (Fu & Kapelan 2013; Wang et al. 2017; Salazar-Galán et al. 2021, Candela et al. 2014; Grimaldi et al. 2012; Grimaldi & Petroselli 2015; Kao & Govindaraju 2007). In these models, empirical distributions are determined based on the results of multiyear rainfall measurements. In the next stage, theoretical distributions that accurately reflect the variability of data in the analysed period are sought. Appropriate statistical tests (Grimaldi & Serinaldi 2006; Sharma et al. 2007; Szeląg et al. 2020; 2022c) are used to assess the compliance of calculation and measurement results. In the next step, to start modelling the time series, it is necessary to determine whether the variables of rainfall characteristics are dependent, i.e. whether they are correlated or not. This is achieved by calculation of Spearman correlations. When the variables are independent (the value of the correlation coefficient has a negligible effect on the simulations), simulations of individual rainfall characteristics are performed using random number generators based on the Monte Carlo method. When the calculated correlation coefficient values are sufficiently large, a multidimensional probability density distribution should be created. This is achieved with the use of cumulative functions, also known as copula functions, since many statistical packages have currently implemented ready-made toolboxes for solving this task (Sadegh et al. 2017; Brechmann & Schepsmeier 2013). An alternative solution to the generation of multidimensional distributions, which require specialist knowledge although they are commonly implemented, is the Iman-Conover (1982) method originally employed for modelling ecological phenomena (Szeląg & Bąk 2017; Wu & Tsang 2004). Computational experiments reported by Szeląg et al. (2018) have demonstrated that this method can also be successfully used to model synthetic rainfall series.

Rainfall generators are useful tools at the stage of modelling and designing sewer networks; hence, they are arousing growing interest. Despite the aforementioned advantages, their use in engineering analyses is limited. One of the main reasons is the limited availability of indispensable software and the need for specialised statistical knowledge.

2.3.3 *Rainfalls with variable intensity*

The use of historical rainfall data is currently relevant, for example, for long-term modelling of rain flow in the stormwater network together with the transported pollution, in the design of stormwater tanks, and reconstruction of large channels. There are several recommendations in the literature for the selection of historical rainfall data. The greatest problem in the selection of historical rainfall data is the insufficient length of records. Experience has shown that approximately only one to two rainfalls per year are suitable for intensity-type simulation purposes. Before proceeding with the construction of the generator, there is a need to assess the suitability of the data for building the model, to determine the rainfall parameters, and to develop the model structure. In these models, at the stage of rainfall simulation, it is crucial to take into account the correlation between the parameters describing rainfall (rainfall duration, rainfall depth, average rainfall intensity, etc.)—see Figure 2.3.

For the purpose of creating rainfall generators in urban catchments, high-resolution rainfall (up to 1 minute) is used. However, this does not limit the use of rainfall data with a resolution of, for example, 15 minutes to build a rainfall generator; everything depends on local conditions

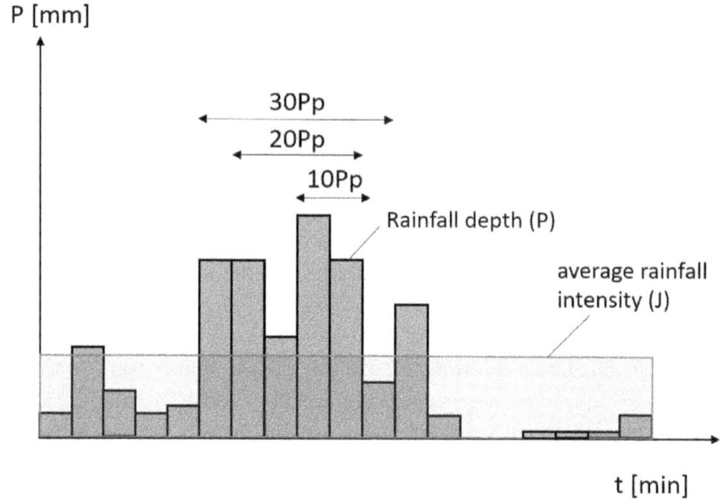

Figure 2.3 Exemplary rainfall parameters for rainfall generator creation.

and the problem to be solved. Despite the resolution of rainfall, the period of rainfall measurement is more important as the input data for the construction of the generator. The literature data (DWA-A 118 2006) show that 30-year rainfall series are recommended for the construction of synthetic rainfall generators. Due to the problems with obtaining such a long rainfall measurement period, it is shortened depending on the problem being solved. On the basis of 10-year rainfall data (measurements with 5-minute resolution), Fu and Butler (2014) showed that, using the method of dome functions, rainfall generators (rainfall duration, rainfall depth) can be developed to model the volume of stormwater overflow in a 200 ha catchment in Scotland. In the paper by Fontanazza et al. (2011), similarly to Fu & Butler (2014), the method of dome functions was used, and a synthetic rainfall generator (rainfall depth and duration) based on 14 years of rainfall observation series was used to simulate the thickness of the stormwater overflow layer in the Centro Storico catchment area (2.5 km^2). Simulations of the continuous operation of the stormwater system with the use of a mechanistic model showed compliance of the calculation results with the data obtained on the basis of synthetic rainfall series. Zhou et al. (2019), based on 30 years of rainfall for the Chengzhong catchment (Jiaxing city), determined synthetic rainfall series using the dome function method, including rainfall duration, rainfall depth, average rainfall intensity, maximum 10-, 20-, 30-minute amount of rainfall in a rainfall episode as input to the regression model for identifying stormwater overflow in the catchment. Szeląg & Bąk (2017) showed that it is possible to use 7-year series of rainfall for the construction of a synthetic rainfall generator, which was confirmed by calculations of the operation of the stormwater overflow. Szeląg et al. (2018) showed that the Iman-Conover method (Iman & Conover 1982) can also be used to build a rainfall generator, the implementation of which is simpler than that of dome functions. In the case of simulating the dynamics of rainfall changes, the requirements for the rainfall measurement period increase significantly. Szeląg et al. (2021c), based on 40 years of rainfall data, developed a model involving the correlation of seven dependent variables to simulate changes in the dynamics of rainfall over many years, which was the next step in the development of rainfall generators.

Another important stage of creating the generator is the separation of independent rainfall events for simulation. Various criteria are used; therefore, in some cases it is crucial that the separated events include heavy rainfall. In other cases when creating a rainfall generator, the specificity of the operation of objects in the stormwater system should be taken into account. In the case of modelling the stormwater runoff, the separated data in the rainfall series should include intense rainfall, which indicates the need to separate events according to the DWA-A 118 (2006) guidelines, in other cases, such as the operation of a stormwater overflow, it should be taken into account that the modelled rainfall events should include whether discharges by transfers have occurred or not. According to the guidelines of DWA-A 118 (2006), the minimum height of rainfall in a rainfall event is 10 mm, and the minimum time between rainfall events is 4 hours, for which the rainfall depth is 2.3 mm, which, in turn. results from the minimum rainfall depth generating surface runoff (Szeląg et al. 2016; Fontanazza et al. 2011).

The use of multidimensional rainfall distributions based on dome functions is a constantly developed scientific aspect. This requires long periods of rainfall measurement. The construction of a two-dimensional rainfall distribution using dome functions is a complex task, and work in this field is continued, with the time distribution of rainfall taken into account (Vernieuwe et al. 2015; Vandenberghe et al. 2011). Attempts are made to correlate the temporal distribution of rainfall with rainfall data (time and height of rainfall) using statistical techniques, which is important from the point of view of reducing the number of empirical rainfall distributions that are included in the rainfall generator. In addition, when modelling the rainfall series, the period between rainfall events and the location of rainfall measuring stations are taken into account (Tarpanelli et al. 2012; Salvadori & De Michele 2006; Eggert et al. 2015). Despite the progress in the construction of rainfall generators, it should be borne in mind that these models are based on probability, and therefore, the precise identification of their occurrence requires further work in this field. The issue of recognising the frequency of rainfall in terms of the season is taken into account in the calculations (Song & Singh 2010; Wong et al. 2010), but the simulation of rainfall in the context of data for RTC systems used to control the stormwater network requires current measurement data collected in the on-line and rainfall forecasts based on mechanistic models to simulate weather conditions.

2.4 Rainfall data requirements

A similar relationship that exists between rainfall intensity and duration, expressed in the fact that high-intensity rainfall lasts a short time, also exists between rainfall intensity and coverage. Short rainfall of high intensity usually covers a small area; a heavy rainfall that covers a larger area and lasts for several hours occurs on very rare occasions. However, in the absence of data on the nature of the movement of the rainfall front, it was assumed that the influence of this parameter on peak flows in the sewer (channel) system is marginal. The issue of spatial variability of rainfall was hinted at as early as the early 20th century, and one of the most commonly cited studies is Specht's formula for determining the ratio of average rainfall intensity I to maximum rainfall intensity I_{max} for a given rainfall area F_{op}:

$$\frac{I}{I_{max}} = \sqrt[12]{\frac{1}{F_{op}}} \tag{2.6}$$

For Polish conditions, the number of publications on this topic is quite modest, and in many cases they concern catchments with large areas. For example, Żyła developed a formula for calculating the rainfall range reduction factor RZ with duration t_d based on a network of 5 rain gauges distributed over a catchment of 2,500 ha:

$$RZ = 1.06\ F^{0.097} t_d^{0.102} \tag{2.7}$$

where: F – catchment area [ha].

However, given the requirements of urban hydrology, where catchments are considerably smaller, this formula has very limited utility. The issue of the extent of torrential rainfall was previously studied by Pomianowski, based on pluviographic data for Warsaw. He established a formula that determines the range of rainfall depending on its maximum intensity (Licznar et al. 2005a,b):

$$X = \frac{75}{\sqrt{I_{max}}} \tag{2.8}$$

where: X – distance from the centre of rainfall [km], I_{max} – intensity at the centre of the rainfall [mm/h].

On the basis of the developed IDFA curves, it was possible to determine the relationship between rainfall duration and range using the Areal Reduction Factor (ARF). The ARF is usually defined as the ratio of the rainfall intensity averaged from n points to the maximum intensity measured, at one point:

$$ARF = \frac{q_n}{q_1} \tag{2.9}$$

where: q_1 – point rainfall intensity, q_n – average rainfall intensity from n points.

For a given rainfall travel speed and rainfall duration, it is possible to determine the range of rainfall at any time t. The average relative rainfall range Z_{op} can be determined for the entire rainfall:

$$Z_{op(t)} = \frac{\sum_{t=1}^{t_{TOT}} F_{op(t)} \cdot t}{t_{TOT} \cdot F_{TOT}} \tag{2.10}$$

where: $F_{op(t)}$ – catchment area covered by rainfall at time t [ha], F – total catchment area [ha], t_{TOT} – duration of rain for the entire catchment area [min].

2.5 Dynamic properties of rainfall

Bearing in mind the results of the research presented in Sections 2.4 and 4.4, an attempt was made to statistically analyse the acquired data in the context of their use in the calculation of sewer networks. This would make it possible, for example, to determine the authoritative dynamic parameters of rainfall for design, taking into account the statistical distribution (probability of occurrence) of the velocities and directions of movement of rainfall cells in a given area.

The analysis included the previously selected 12 rainfall events and an additional 18 events with slightly lower intensities. Due to the relatively low density of the measurement network in question, it was impossible to determine the actual extent and shape of the moving rainfall cells.

A fundamental problem in the analysis was the adoption of a methodology for determining the direction of rainfall movement, which was affected by the high temporal and spatial variability of individual events. Finally, cross-correlation was used for all possible combinations of pairs of rain gauges (10 pairs in total). For each pair of rain gauges (i, j), Pearson 's linear correlation coefficient was determined, defined as:

$$R_{i,j,\tau} = \frac{\sum_{t=1}^{T-\tau} \left(I_{i,t} - \overline{I_i}\right) \cdot \left(I_{j,t+\tau} - \overline{I_j}\right)}{\sqrt{\left(\sum_{t=1}^{T} \left(I_{i,t} - \overline{I_i}\right)^2\right)\left(\sum_{t=1}^{T} \left(I_{j,t} - \overline{I_j}\right)^2\right)}} \tag{2.11}$$

where: τ – delay time, i.e. the time it takes to travel the distance between the two selected rain gauges (i, j), $\overline{I_i}, \overline{I_j}$ – the average rainfall intensity at the i-th and j-th pluviographs:

$$\overline{I_i} = \frac{1}{T}\sum_{t=1}^{T} I_{i,t}$$

$$\overline{I_j} = \frac{1}{T}\sum_{t=1}^{T} I_{j,t} \tag{2.12}$$

The relative position coordinates of rain gauges i and j were introduced:

$$x_{i,j} = x_i - x_j$$

$$y_{i,j} = y_i - y_j \tag{2.13}$$

where: x_i, y_i – actual coordinates of the pluviograph i, x_j, y_j – actual coordinates of the pluviograph j.

With this assumption, the cross-correlation does not depend on the absolute location of the rain gauges, but only on their relative location. Therefore, the cross-correlation between two rain gauges will be a function of three variables: τ, and $y_{i,j}$. In the relative datum, each pair of rain gauges gives two points, corresponding to the values of the correlation coefficients $R_{i,j,\tau}$ and $R_{j,i,\tau}$, the coordinates of which are: $(x_{i,j}, y_{i,j})$ as well as $(-x_{i,j}, -y_{i,j})$, respectively.

The range of obtained rainfall movement velocities (3–12.5 m/s) indicates that these will practically always be higher than those observed in the sewer system. Given that the greatest impact on the peak values occurred with rainfall moving at a velocity of 2 m/s, the probability of such a case is small. In the case of Czestochowa, rainfall moving at 5 m/s in an easterly direction should be considered likely.

If data from several pluviographs are available for a given catchment area, it can be proposed to determine the weighted average h_{OpS} rainfall, where the weight will be the catchment area representative of a given pluviograph:

$$h_{OpS} = \frac{h_{Op1}F_1 + h_{Op2}F_2 + \cdots + h_{Opn}F_n}{F_1 + F_2 + \cdots + F_n} \tag{2.14}$$

where: h_{Op1}, h_{Op2}, h_{Opn} – amount of rainfall recorded on pluviographs, F_1, F_2, F_n – areas of individual catchments.

The formula above makes it possible to determine the average total rainfall in a given area, i.e., it accounts for spatial non-uniformity, but does not account for temporal non-uniformity.

Determining the characteristics of rainfall, while accounting for its spatial variability, will additionally require the determination of the so-called surrogate histogram (Figure 2.4). Again, a weighted average can be used, but this average must be calculated for each subtotal of rainfall recorded over the time interval t_i at all n rain gauges:

$$h_{Op(t_i)} = \frac{h_{Op1(t_i)}F_1 + h_{Op2(t_i)}F_2 + \cdots + h_{Opn(t_i)}F_n}{F_1 + F_2 + \cdots + F_n} \tag{2.15}$$

where designations are the same as above.

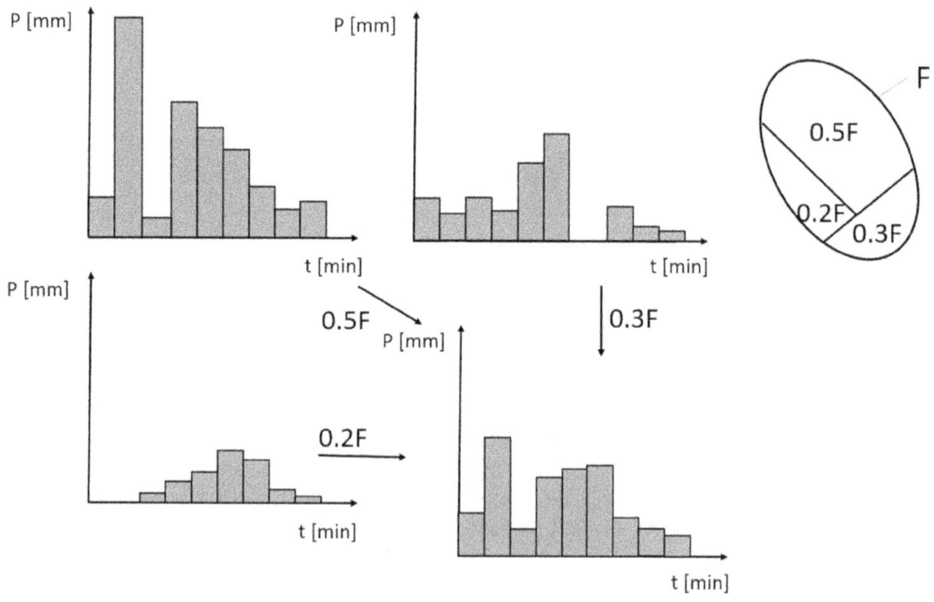

Figure 2.4 Schematic for determining a surrogate rainfall histogram for a catchment area of *F* (based on Mrowiec 2009).

2.6 Urban catchment data

The data on the drained area significantly influence the results of the calculations and should, therefore, be obtained accurately, if necessary also in detail and predictably (especially with regard to the future of the drained area). From the drainage-technical point of view, not only is the extensive expansion of the drained area meant, but also the regulatory measures in the river basin, which are aimed at reducing the amount of runoff and pollution from the river basin. The detail of determining the necessary data on the river basin depends to a large extent on the type of task, and thus also on the requirements of the rainfall-runoff model used.

For developing a combined sewer network model, it is necessary to collect data on both the characteristics of the catchment and the stormwater network as well as quantitative data indicating inflow from sanitary and industrial sewers. For this purpose, it is necessary to measure the amount of stormwater outflow on a daily basis with an hourly resolution, so that its typical variability can be determined for hydrodynamic models. Due to the price of samplers used to measure the amount of wastewater and the scattered development, their number—and thus the cost of metering the system – would be high. Therefore, at the stage of developing hydrodynamic catchment models, continuous measurement results from water metres are used. By introducing a reduction factor, it is possible to approximate the amount of stormwater flowing into the network.

The required data are data decisive for surface runoff, but area-related data that are decisive for runoff in the dry weather period season. The main data for determining the surface runoff from the urban area are:

- total catchment area,
- drained area,
- subcatchments of the sewerage network,
- impervious, drainage areas,
- pervious areas and their characteristics, influencing the outflow from them,
- slope of the particular subcatchments.

Delineation of sub-catchments was discussed in detail in the work of Szeląg et al. (2022b), landform and its delineation were discussed in detail in Szeląg et al. (2016).

Careful identification of catchment area data is essential for tasks with the required high level of detail. For this, e.g. evaluation of aerial photographs, cadastral maps, inspection around the territory, etc., is required. These approaches are recommended, if not for the whole territory, then at least for representative sub-areas.

Regardless of the tool used to model catchment run-off, it is important to identify the impervious areas that have a large impact on the surface run-off formation. An impervious area is defined as an area that prevents or delays run-off to the ground compared to impervious area (Burszta-Adamiak 2014). In addition to determining the size of the impervious area for modelling, the manner the area is connected to the stormwater network model is also of great importance (Zoppou 2001). Several cases should be kept in mind, the first one being when the stormwater from the sanitary system, the development and additionally the paved surface under which the run-off is discharged into the stormwater network. When creating a catchment model, the cases where the stormwater from the roof directly falls onto an impervious area should also be taken into account. This fact has to be taken into account in the model, as omitting it will affect the values of the calibrated parameters and the uncertainty of the simulation results,

Table 2.2 Dependencies for determining the value of effective impervious area surface connected directly to the stormwater network

Catchment characteristics	*Equation*
Catchment area with impervious areas mostly disconnected from the stormwater network	$0.01 \cdot Imp^{2.0}$
Impervious areas partially connected to stormwater system, roof run-off used locally	$0.04 \cdot Imp^{1.7}$
Average connection conditions of impervious areas (roofs disconnected from the sewerage system)	$0.1 \cdot Imp^{1.5}$
Impervious areas (including roofs) connected to a stormwater network to a high degree	$0.4 \cdot Imp^{1.2}$
Impervious surfaces directly connected to stormwater network	Imp

which have great influence on the reliability of the obtained simulation results. Bearing the above-mentioned problems in mind, appropriate dependencies were given for determining the value of the so-called effective impervious area surface connected directly to the stormwater network (Mrowiec 2009) – see Table 2.2.

On impervious paved areas, attention must be paid to the determination of the roughness for the model expression of surface runoff, the type of surfaces in the river basin. For detailed modelling, separate determination of roof surfaces and other impervious surfaces is suitable, depending on their drainage characteristics and slopes.

In particular, it should be considered whether there is an outflow from the area into the sewer network. E.g. if the road does not have stormwater drainage facilities, runoff to the sewer network will be minimal. Consideration should also be given to whether stormwater runoff will continue to be connected to the public sewer or whether rainwater in the future will be infiltrated.

Depending on the mathematical model of the runoff, data on the type of substrate and the vegetation cover, the size of the slope and the direction of the slope, the facts preventing the runoff and forming the runoff are also important for the simulation of surface runoff from pervious surfaces.

The most important data for determining the dry weather (sanitary) flow are the number of connected inhabitants, specific wastewater production from the population, population density, daily amount of wastewater, factor to take into account fluctuations in water runoff during the day, external water runoff, industrial water runoff, and other inflows.

It is recommended to confirm determined runoff quantities and their daily and annual fluctuations by measurements.

Data about the technical condition of the stormwater network are also important; the capacity of a stormwater network is a factor directly attributable to the condition of the sewers. Leaky, displaced joints, deposited sediment at the bottom, and damaged sewer construction are direct factors that have a large impact on stormwater capacity. Diverse hydraulic conditions in pipes during the stormwater flow, due to the transport of mineral material, generate varying resistance at the wetted perimeter, which contributes to abrasion of the pipe walls and reduces the pipe thickness. This, in turn, reduces the load capacity of the pipe structure as well as increases the roughness coefficient of the pipe. It is undeniable that these analyses for individual sewers are complicated, and even more so for the entire stormwater network. This is due to the fact that there are numerous hydraulic phenomena occurring in a stormwater network (backwater, backflow, hydraulic jumps, aeration of the sewage stream due to increase of the sewer slope, in cascade wells, etc.) which translate into the arrangement of stormwater table in pipes and the determination of hydraulic load of individual pipes operating without pressure and with complete filling.

2.7 Sewer network system data

Drainage systems in urban areas consist of a large number of different elements that are interconnected. Mathematical models of runoff describe the interaction of these elements through hydraulic and hydrological dependencies. The interaction of the elements presupposes their geometric and hydraulic definition and determination of their interconnection (network topology) in the digital model of the drainage system.

Parts of the drainage system are divided on the basis of hydraulic functions into:

- transport elements, i.e. open or closed conduits such as underground pipes, surface gutters and drains, pressure pipes, ditches,
- point elements—manholes, confluence manholes, storage facilities, tanks, etc.,
- distribution devices such as overflows, combined sewer overflow chambers,
- control and regulating devices such as shutters, check valves, throttling elements, pumping stations, etc.

Transport elements are generally determined by the start and end point. These are usually the manholes at the beginning and end of the section. They can also be nodes in the network without manholes (virtual nodes), objects, or control elements. For the mathematical simulation of the runoff, data on the element type (open, closed sewer pipe, pressure pipe), topological definition (start and end point), geometric characteristics, (e.g., section length, which can be indirectly defined by the coordinates of the start and end point), invert and ground levels, profile shape and dimensions, hydraulic roughness, local losses, etc.) are also expected.

Point elements do not require a topological definition; usually, a geometric definition is enough to describe them. In general, data on the type of element (manhole, tank, outlet, etc.), geometric characteristics (diameter, dimensions, coordinates, ground, invert level, etc.), and hydraulic characteristics (local losses, in accumulation dependence of flow area and volume on water level) are directly related to facilities with specific topology in urban catchments.

Distribution devices distribute the inflowing water into several transport elements. At the same time, they may be able to partially accumulate water. While data on the height of transport elements are sufficient for pipe and duct branches, additional information on the element type (distribution chamber, CSO chamber with frontal or side overflow), geometric characteristics (dimensions, coordinates, dimension, crest level, length and shape of the crest, etc.), ground and invert level, and hydraulic characteristics (local losses, overflow coefficient) are expected.

Control devices are usually classified as special devices. To describe their hydraulic effect, the type of equipment, hydraulic characteristics, and dimensions must be entered; the type of element (valve, non-return valve, flow regulation, pump, etc.), geometric characteristics (dimensions, coordinates, dimension, length and shape of the overflow edge, etc.), ground and invert level, mode of operation (method of equipment regulation, e.g. for pumps switching on, switching off levels, etc.), and hydraulic characteristics (local losses, coefficients).

2.8 Data for water quality modelling

2.8.1 *Characteristics of stormwater runoff*

It is difficult to qualitatively assess stormwater runoffs, although they can be characterised by specifying, for specific cases, the content of various groups of contaminants. These groups include (Królikowska & Królikowski 2012):

- settling and non-settling suspension, as well as larger solids particles,
- inorganic substances,
- organic matter,
- living organisms, especially pathogens, bacteria and viruses, protozoa, and small tissue organisms.

The composition of polluting substances resulting from stormwater runoffs and contaminants occurring in them is typical for given catchment and atmospheric rainfall. Proper and accurate stormwater runoff assessment can only be made based on the results of direct testing cycle of runoffs from a specific catchment. They are reliable only for the catchment and for the case of a given range of parameters (Price & Vojinovic 2011).

The composition of stormwater runoffs can be described using similar indicators as in the case of municipal wastewater. In that case, the following are taken into account (Szeląg et al. 2013; Królikowska & Królikowski 2012):

- concentration of total suspension/total suspended solids – TSS,
- biochemical oxygen demand – BOD,
- chemical oxygen demand – COD,
- concentration of biogenic substances,
- concentration of heavy metals,
- concentration of chlorides,
- concentration of petroleum derivative substances.

Suspension concentration is the most significant factor of sewage contaminants caused by atmospheric rainfall. The problem is constituted primarily by contaminants associated with suspension, such as organic compounds, heavy metals, bacteria, and viruses (Table 2.3). Additionally, a significant share of oil contaminants and the phenomena occurring on the surface of suspensions complicate the issue (Marsalek et al. 1998). They are, for instance, ion exchange,

Table 2.3 Contaminants accumulated in suspension (Królikowska & Królikowski 2012)

Contaminants	*Percentage of contaminants associated with the suspension*
COD	83%–92%
BOD	90%–95%
Total nitrogen	65%–80%
Hydrocarbons	82%–99%
Lead	97%–99%

catalysis involving enzymes and mineral catalysts. In stormwater runoffs, mineral suspension is a predominating element. Organic suspension, on the other hand, occurs in catchments like market places, green areas, especially in the autumn as a result of falling leaves. Size distribution of particles appearing in stormwater of urbanised areas is distinguished by high changeability and it constitutes an individual characteristic of a catchment. In most of the analyses, it is concluded that the largest share is that of particles in the range of 10–300 μm.

The suspension concentration in stormwater runoff varies significantly and often exceeds values typical for sanitary sewage. It depends on the catchment management, the type and degree of surface imperviousness, traffic intensity, rainfall parameters, and the number of dry days before the rainfall, and also on the sewer exploitation (Muleta et al. 2013). A high concentration is observed in the runoff of disordered catchments where substrate erosion from unsealed surface occurs. During a downpour, runoff water grabs even the heaviest suspension fractions with high weight, which allows their easy sedimentation. Over the duration of rainfall, the share of suspension in stormwater runoff decreases. The share of the organic suspension increases in the case of green catchments in autumn. This takes place due to the presence of leaves and other dead plant matter.

A significant concentration of suspended solids appears also in snowmelt runoff and often exceeds their concentration in stormwater runoffs (Linsley et al. 1958). This suspension is primarily mineral in nature, especially when, apart from chemical substances (usually chlorides), sand and other mineral aggregates are used to prevent the formation of glazed frost (Królikowska. & Królikowski 2012). In this case, for the most part, the suspension is combined with various kinds of organic and inorganic contaminants.

The individual size of suspension fractions is significant in the transport of contaminants integrated with it. The smallest fractions contain the highest contaminants percentage, as they have the most developed surface, which has an impact on sorption properties of molecules (Table 2.4).

The value of biochemical oxygen demand (BOD) in stormwater is usually low and often does not reflect the actual amount of organic compounds. This is because biochemical processes are usually hampered by toxic substances and inhibitors present in stormwater (Dąbrowski 2004).

In the runoffs from roads, car parks and manoeuvring areas, BOD is not a good indicator to determine the qualitative characteristics of stormwater runoffs (Dąbrowski 2004). Because of these reasons and due to the presence of petroleum derivatives in stormwater, the COD indicator is much more convenient (Table 2.5).

Chemical oxygen demand (COD), defined as the amount of oxidised substances in a chemical way, usually characterises the degree of stormwater runoff contamination quite well.

Table 2.4 Contaminants percentage associated with individual fractions of suspensions in stormwater runoffs (Królikowska & Królikowski 2012)

Total suspended solid value	*Percentage of total suspended solid*	*Contaminants percentage associated with individual fractions of suspensions*			
		COD	*BOD*	*Total nitrogen*	*Lead*
> 250 μm	11–40	30	25	13	41
50–250 μm	6	0	9	3	41
10–50 μm	10–20	12	14	11	3
< 10 μm	45–63	58	52	73	56

Table 2.5 Approximate contaminants in stormwater runoffs (Królikowska & Królikowski 2012)

Contaminants indicators	*Contaminants values in stormwater runoffs*	
	Minimum and maximum values	*Average values range*
Total suspended solids (g/m^3)	5–40,000	80–17,000
BOD (g/m^3)	1–1,100	7–83
COD (g/m^3)	5–3,100	40–50
Chlorines (g/m^3)	5–428	12
Total nitrogen (g/m^3)	0.3–12.7	3–10
Lead (g/m^3)	0.9–12.6	–
pH	3.9–8.7	5–8

Moreover, there are often dependencies between the COD value and the concentration of other contaminants, such as heavy metals. COD value varies from several tens to even several thousand mg/dm^3, and often exceeds the value of 150 mg/dm^3 (Królikowska. & Królikowski 2012). COD and its fractions have been characterised in detail in the subsection concerning sanitary wastewater parameters.

In stormwater runoffs, the concentration of heavy metals is also significant, mainly due to the possible presence of lead. Nevertheless, since the introduction of unleaded petrol on a wider scale, the problem in Poland and the European Union is now less important than in previous times. Heavy metals are mainly present in the runoffs of stormwater from traffic routes, as well as heavily urbanised and industrialised areas (Kotowski 2011). They come from unburned fuel fractions, wear of tyres and asphalt. In most cases, they are toxic and carcinogenic compounds. Heavy metals may be present in wastewater as dissolved salts, ions, or insoluble compounds (Królikowska & Królikowski 2012).

Chlorides occur generally in snowmelt runoffs as a result of using chemicals to prevent the formation of glazed frost. These are mixtures of NaCl, $CaCl_2$, and sometimes $MgCl_2$.

Petroleum derivatives present in stormwater runoffs are all organic substances from the use and processing of crude oil. They are present in wastewater in dissolved form, as an emulsion, or in suspension. Stormwater runoffs can also be contaminated with aromatic hydrocarbons derived from unburned fuel fractions or remaining as products of tyre wear and asphalt surface. Higher concentrations of petroleum derivatives and oils in stormwater runoffs outside petrol stations may be the result of accidents and road disasters, as well as the use of faulty seals in drive mechanisms, or of hydraulic leaks in motor vehicles.

The pH of stormwater runoffs is usually neutral. However, while rain has contact with air containing carbon dioxide and other acidic substances with clearly increased rainfall within many urban areas, rainfall pH value decreases below 7 (Imhoff & Imhoff 1996). In turn, on the catchment surface alkaline substances may occur. Thus, pH of stormwater runoffs flow can be in the range of 5–8, and sometimes even 9.

2.8.2 *Sediments and their transport in gravitational sewer systems*

Sediments in gravitational sewer systems appear when the velocity of flow decreases below the value of hydraulic self-purification. They have a significant influence on the processes occurring in sewer systems as a cause of, e.g. periodical fluctuation of pollution loads entering wastewater treatment plants. Sediments deposited during the phase of low canal filling and low flow velocity reduce the pollutant load; thus, during the phase of high filling and high flow velocity, the deposited pollutants are washed away, thereby rapidly increasing the concentrations and loads of transported pollutants. This phenomenon directly influences the pollutant load entering the wastewater treatment plant or storm overflow and the sewage discharge receiver. Sediments deposited in the channel also cause changes in the pipe cross-section shape and thus bring about changes in the physical and hydrodynamic parameters of the entire system (Hirt & Nichols 1981; Geiger et al. 1987; Dąbrowski 2004; Hager 2010). Sediments also contribute to the development of sulfuric corrosion of pipes and have an impact on the intensity of biological biodegradation of wastewater in sewer conduits.

The techniques of sampling and measurements applied currently yield results that only partially reflect the properties of sediments in their natural environment.

Since studies concerning wastewater sediments and related processes are relatively expensive, they are only conducted in countries in which the biding law requires mathematical modelling of mass transport to wastewater plants and storm spillways (Dąbrowski 2004). The results of this research provide data necessary to calibration of hydraulic models.

Classification of sediment forms in sewer systems.

The following classification of sediments is used (Marsalek 1998; Hvitved-Jacobsen 2002; Dąbrowski 2004):

A non-organic and non-cohesive sediments,
B sediments with the A-type non-organic structure but agglomerated by oil derivatives and characterised by cohesive properties,
C organic sediments, moving over the surface of sediments A and B with lower velocity than flowing sewages,
D biofilm developed on pipe walls,
E sediments in storage tanks.

Two other groups of sediments are sometimes distinguished (Dąbrowski 2004):

F storm conduit sediments,
G pressure pipe sediments (after pressure sewer attachment).

Sediments A and B are the major types found in wastewater systems, thus, they exhibit the highest load of COD accumulated, although the concentration of pollutants in these sediments is lower than in the case of type C.

The highest content of organic pollutants has been noted in sediments from group C, in which the COD content reaches the level of up to several hundred (even 300–500 gO_2dm^{-3}). These sediments move over the surface of group A and B during the phase of low sewage flow in a wastewater pipe. The presence of sediments C and D in sewer systems results in occurrence of biodegradation processes, because these sediments contain heterotrophic organisms, the process factor of the transformations mentioned.

Group D sediments comprise heterotrophic living organisms; therefore, the organic content in this group of sediments is high. The thickness of the layer formed by group D is usually low (generally not exceeding 2 mm) but the anaerobic conditions are possible. This, in turn, contributes to formation of hydrogen sulphide, which induces sulfuric corrosion of pipe surface, subsequently attacking the vault and side walls of the canal.

Group E sediments are characterised by a high degree of fragmentation; hence, the grain surface area in the unit volume of sewage is significant. This facilitates extensive adsorption of practically insoluble organic pollutants and heavy metals.

Group F sediments present in wastewater and storm systems are characterised by variable, time-dependent (seasonal) composition. In particular, they contain large amounts of leaves falling from trees during late autumn. Pressure pipe sediments (group G) commonly occur at the pipe slope and contain huge amounts of organic compounds, especially fats.

On the basis of measurements, including the mechanical properties of sediments, it was found that critical shear stresses τ_y depend on the density of analysed sediments in their natural environment (Price & Vojinovic 2011). Polish studies (Dąbrowski 2004) have demonstrated that the phenomena connected to sediment transport and pollutant release are largely dependent on the relations among particular sediment properties, the most important of which include density, moisture, and porosity. In Slovakia, the critical value of the shear stress for the self-cleaning ability of the channels is set by the standards with the value of 4 Pa. Another possibility to maintain the self-cleaning ability of pipes is to design them with a so-called self-cleaning slope (approx. $I=3{,}000/DN$, where the DN is the pipe diameter in mm, i is the slope in ‰, one-per-thousand).

2.8.3 Receiving water data

In certain cases, quantitative and qualitative data on recipients may also be needed when modelling runoff from urbanised river basins. These data are mostly needed modelling various complex scenarios of water quality development, e.g. long-term balances of pollution transport into watercourses, short-term impact on quality or flows in the recipient (hydraulic or chemical stress) due to hydrological events in the urbanised river basin (storms) etc. These data may also be needed in complex modelling of urban flooding, assuming that the sewerage network and the surface flow network form a complex unit.

The quantitative data needed for sewage networks, and WWTPs assessment include hydrological and hydraulic data, e.g. data on river basin catchment, water levels, water flow rates. The stated data are usually measured and recorded at the water meter stations of the Hydro Meteorological Service of the relevant state (e.g. in Slovakia, SHMÚ). The Hydro Meteorological service archives and evaluates these data, so that not only metadata are usually available (e.g. lists of water meter stations), but also data on average and extreme water levels and flows (monthly and annual averages, significant peak flows, minimum flows, daily flows, etc.). Sometimes the availability of this data can be a problem due to payments required for providing them for commercial purposes (for study and scientific purposes, they can be provided free of charge).

Similarly, the Hydro Meteorological Service of the particular state typically also evaluates the quality of water in watercourses. However, it should be noted that the time range of most of the monitored parameters is significantly lower than that of the quantitative data. While water levels and flows can be monitored and recorded at minute intervals, qualitative data are available at significantly longer intervals due to the cost of sample collection and chemical analyses. Water quality checks are usually performed once a year (small watercourses without significant water quality problems), monthly (water quality control profiles on most streams), or daily (large international watercourses).

The Hydro Meteorological Service performs the collection and chemical analysis of surface water samples, as well as the assessment of water status according to national legislation based on the European Directive 2000/60/EC (https://eur-lex.europa.eu). The data are typically processed in the form of annual reports. These reports, in addition to qualitative data and their summary evaluation (assessment of the overall water status), also contain lists of major pollution producers and monitoring results of some special substances (priority substances, hazardous substances, micropollutants, etc.). The annual reports also contain a list of river profiles in which good water status has not been achieved (a goal for all EU Member States) and the reason for not achieving this good water status. In addition to qualitative data, the annual reports may also contain data on the characteristic flows Q355, Q270, as well as the long-term average annual flow Q_a and the annual flow Q1.

The Hydro Meteorological Service should annually review the requirements for water quality that result from the objectives of surface water quality protection in individual river basins. The monitoring network focuses on the sections affected by discharged pollution, on the identification of progressing pollution, on the evaluation of quality development trends, as well as on the provision of indicative data to assess the suitability of water for further use.

Another source of data may be the river basin Hydro Ecological Plans (HEPs) prepared by the river basin authority or the relevant state authorities. The HEP section on water quality development and forecast contains data on point and area sources of pollution, surface water quality, data on sediments, and groundwater quality. Longitudinal pollution profiles for prescribed water quality parameters are important information as well. From the graphical evaluation of changes in the water quality of the stream in the form of a longitudinal profile, one can see the course and trends of water pollution.

2.9 Geographic information systems and remote sensing

In practice, designers and operators often run into problems when it comes to obtaining up-to-date information on existing networks or obtaining hydrological and other data from a given area. The accuracy of these data determines the efficiency and investment of the designed sewers or the possible rehabilitation of existing sewers. In operation, therefore, the so-called sewerage information systems (often referred to as GIS –geographic information system) are often encountered, which have to eliminate the time-consuming work of obtaining information and, at the same, time interconnect large amounts of data according to the context.

GIS is a tool for computer collection, manipulation, processing, analysis, and display of spatially oriented data. This system contains geometric data (coordinates, points, lines, areas, and topological information) and their attributes, i.e. information describing the properties of geometric objects. Attributes can be not only alphanumeric but also graphical information about objects, e.g. cross-sections, photographs, etc. GIS contains tools for manipulating both types of data: graphic data (graphic editor), as well as attributes (database, graphic editor, etc., according to the nature of attributes) and tools for creating links between both types of data.

The basic data that every designer and operator should obtain through this system include geometric, construction, hydraulic, operational and economic data on the sewer network. At present, three basic levels of information systems are distinguished according to their complexity:

1 Digital map (graphical record) and / or alphanumeric database (without interconnection),
2 Graphic information system (digital graphic + alphanumeric record with interconnection),
3 Comprehensive GIS-sewerage network data are only a subsystem of the overall and complex information system of the city (region, etc.) and are interconnected with other data subsystems.

The third level of complex GIS seems to be the optimal solution. The user can have access, e.g., to information on other supply networks, on current property owners (based on data from the real estate cadastre), etc. However, such a system requires the cooperation of several institutions that collect and manage such data. This brings not only technical problems (data models, interconnection, etc.), but also the problems related to copyright, use, or confidentiality of certain data. Because of this, information systems which can be included in the second level of GIS complexity are currently mostly encountered.

It is also important to connect the rainfall-runoff model with the information system on both sides, so that the simulation results can be supplemented as data in the information system (e.g. maximum levels, capacity utilisation of collectors during rainfall, etc.).

An interesting topic connected with the use of GIS is remote sensing. For many years, an increased integration of GIS technologies, including remote sensing techniques, with hydrologic modelling has been observed. This is supported by an increasing amount of spatial data in the form of numerical terrain and land cover models, land use maps, soil classification, and contour maps, which are available to hydrology researchers. In order to conduct hydrologic modelling, it is essential to provide accurate spatial data and to understand its impact on the physical processes occurring in the environment and underlying the hydrologic cycle.

Remote sensing images can provide extensive spatial information as a function of time. They are useful during parameterisation and calibration of hydrological models. Among the many applications of remote sensing techniques in environmental engineering, one can mention the quantitative determination of the surface of different types of land use, surface roughness, extent of snow cover and flooded areas, as well as the qualitative monitoring of inland waters and soils in agricultural areas. An additional advantage of remote sensing techniques, as opposed to in situ measurements, is the possibility of verifying the results of spatial models and parameters relevant to hydrological modelling in the form of moisture and Leaf Area Index (LAI) values, ultimately used in the calculation of evapotranspiration in the water balance of the entire analysed catchment (Chormański et al. 2008).

An important parameter for spatially varying hydrologic patterns in urban catchments is the degree of catchment imperviousness, otherwise known as the catchment impervious area. The volume of surface run-off increases along with impervious area, which hinders aquifer recharge and accelerates soil erosion. As a result, local flooding and traffic obstructions may occur. The value of impervious areas for an urban catchment increasingly often serves as an indicator of the ecological status.

The methods of identifying impervious areas can be divided into two groups, i.e., direct and indirect methods. Indirect methods involve fragmenting a large land area and separating general land cover classes within it. Small-scale land cover and land use maps are publicly available for the EU countries in connection with the EU CORINE LULC project. Each land cover class is percentage-related to the impervious area. Due to the generalisation of land cover types and the lack of standardisation of the method in question, there is a risk of large variations in the percentage of impervious surface within the same land cover type; therefore, indirect methods only allow an approximate estimation of the amount of impervious surface for a catchment. For a more accurate determination of the impervious surface, direct methods are used, based on orthorectified aerial images subjected to subsequent mosaicking, industry data in the form of base maps and databases of topographic objects, and multispectral satellite images (Chormański et al. 2008).

The identification and subsequent vectorisation of impervious areas on the basis of a high-resolution ortho-photo-map constitutes one of the most accurate techniques used for this purpose. Due to the necessity of manual identification of objects and defining their outline, this method is time-consuming and rarely used in practice, despite its great potential. The use of sector data available in local government units is a simplification; however, the data are not always up to date for the year of analysis, which is important in times of increasing urbanisation.

Satellite images and supervised classification of their content allow quick and reliable determination not only of the area but also the distribution of impervious surfaces, even for very large areas, which is advantageous for sub-basin analyses (van der Sande et al. 2003). Another well-known approach to the separation of built-up areas from ecologically active areas is thresholding of selected spectral indices, calculated on the basis of recorded electromagnetic radiation in successive spectral channels (As-syakur et al. 2012). The examples of such indices are:

- NDVI – calculated from spectral channels recording red and near-infrared radiation. The lack of correlation between these channels is highlighted by the differentiation of urbanised areas from vegetation that absorbs red radiation and reflects infrared radiation,
- NDBI – enables to distinguish the developed from unvegetated land,
- LAI – A leaf coverage index that determines how much light is used by plants. The higher the value, the more energy is used for photosynthesis.

Chapter 3

Urban runoff calculation methods

3.1 Rational method

One of the approaches for the estimation of maximum flow values and duration of design rainfall in small urbanised catchments is the constant rainfall intensity method (CRI). In this approach, the calculations are based on the assumption of constant rainfall intensity with assumed rainfall frequency related to area drainage and duration. This assumption is typical of hydrological methods used in urban catchments, which is due to the lack of rainfall measurements. Important parameters of an urban catchment include the flow coefficient connected with surface pavement and the kind of housing (Figure 3.1).

The CRI method assumes that the rainfall duration is equal to the channel flow duration and is calculated with the following formula:

$$t_p = \frac{L}{\upsilon} \tag{3.1}$$

where: L – channel length (m), υ - channel flow velocity (m/s).

As shown by literature data (ATV-A 118 1999), the shortest duration of design rainfall used for calculations is 10, eventually 15 minute, and the value of the maximum flow is defined as:

$$Q = \phi(s) \cdot \psi \cdot A \cdot q(t_r) = \phi(s) \cdot q(t_r) \cdot A_{zr} \tag{3.2}$$

where: $\phi(s)$ – outflow delay coefficient, ψ – flow coefficient.

The area of the substitution catchment determining the flow volume, also referred to as the reduced catchment, is calculated as follows

$$A_{zr} = \sum_{j=1}^{m} A_j \cdot \psi_j \tag{3.3}$$

where: A_j – area of the jth subcatchment (ha), ψ_j – flow coefficient of the jth catchment.

The flow delay coefficient shown in Eq. 3.2 is also described by the Bürkli-Ziegler equation (Cinger 1960):

$$\phi(s) = A^{-1/s} \tag{3.4}$$

where: s – empirical coefficient describing flow condition in the system based on the velocity.

DOI: 10.1201/9781003546436-3

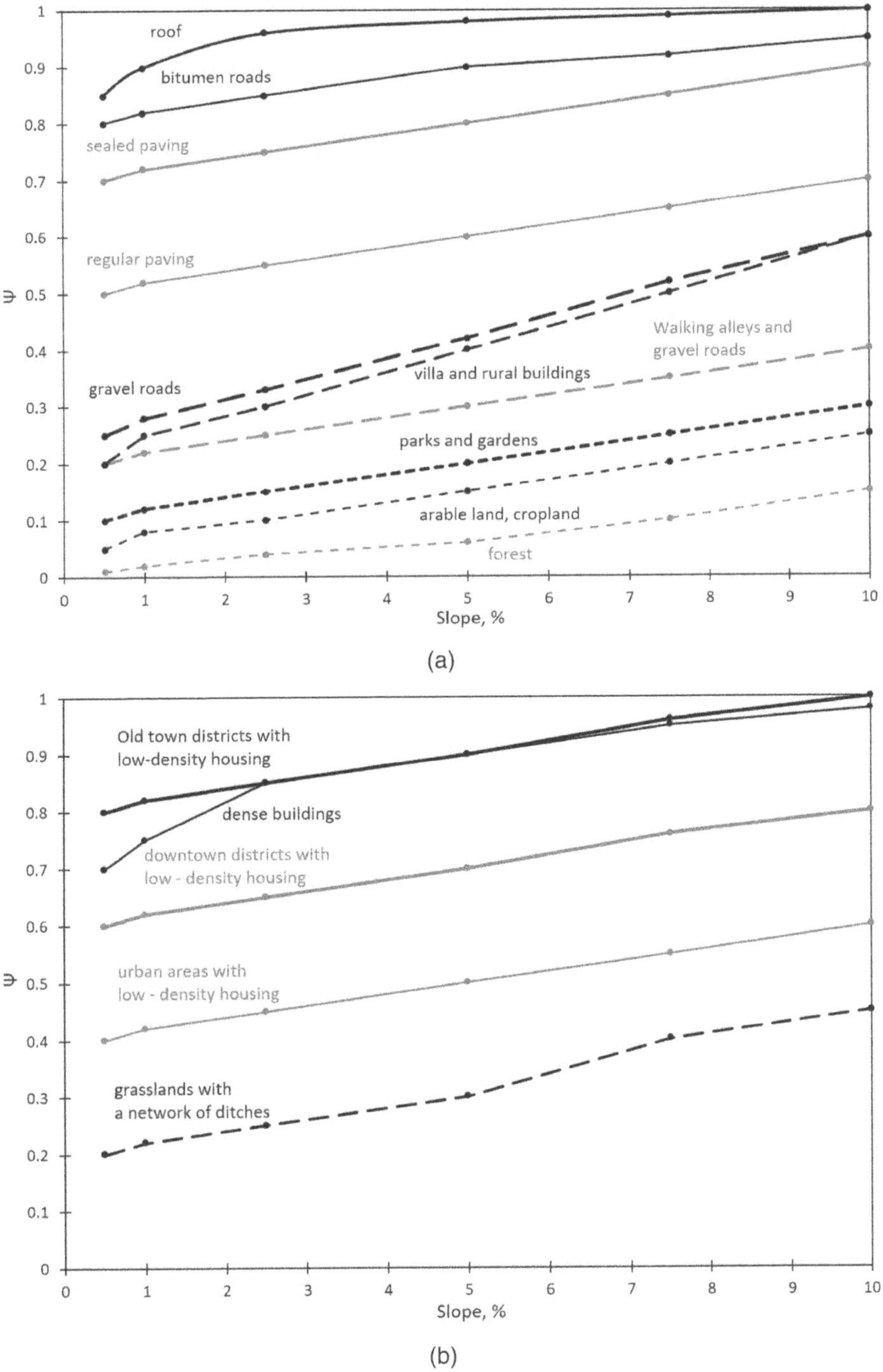

Figure 3.1 Relationship of the runoff coefficient with terrain slope and land use connected with pavement (upper graph) and connected with kind of housing (lower graph). (Based on the data provided by Ciepielowski & Dąbkowski 2006.)

The s coefficient in Eq. 3.4 takes into account the non-simultaneous stormwater outflow from the catchment to the computational cross-section closing the catchment. The value of coefficient φ takes into account the scheme of the sewer system, channel and terrain slopes, and the hydraulic roughness of the runoff surface; however, this relation has an approximate character. Under moderate conditions, when the terrain and channel slope determine the flow velocity of 1.2 m/s in the channel, and when the length of the catchment area is twice as large as its width, the value $s=6$ is assumed. The value $s=4$ is taken in the case of more centred catchments with higher slopes (Kotowski 2011).

3.2 Threshold rainfall intensity (TRI) method

This method is one of the frequently employed approaches in designing stormwater sewer systems derived from the rational method, as mentioned earlier. In TRI calculations for a sewer network, the time of design rainfall (t_{dm}) is determined for each node connecting two channels (connection manhole, straight-through manhole), taking into account rainfall frequency (C) and rainfall duration. The duration of design rainfall is calculated with the following formula (Błaszczyk et al. 1974):

$$t_{dm} = t_k + t_p + t_r \tag{3.5}$$

where: t_{dm} – duration of design rainfall (min), t_k – time of concentration of the surface runoff (min), t_p – channel flow time (min), t_r – channel retention time (min) established at $t_r=0.2t_p$.

Given these relationships, the following dependency can be inferred:

$$t_{dm} = t_k + 1.2 \cdot t_p \tag{3.6}$$

In the TRI method, the minimal design rainfall duration has the same value as in the CRI approach, i.e. 10 minute (Błaszczyk et al. 1974; SNiP II 1976; Mrowiec 2009).

The time of concentration (t_k) is the time required for water to flow from the most remote point of the catchment to the channel inlet. Its value depends on the intensity of rainfall, the type of surface over which stormwater flows, the distance from the inlet to the most remotely located point in the catchment, and the terrain slope (Akan & Houghtalen 2003). In the threshold rainfall intensity approach, the time of concentration is determined based on rainfall frequency, surface type, and average slope (Table 3.1) (PN-S-02204 1997; PN-EN 752-4 2001; ATV-A 118 1999).

Table 3.1 Time of concentration and rainfall frequency (PN-S-02204 1997; PN-EN 752-4 2001; ATV-A 118 1999)

Type and location of the channel	*C* / *years*	t_k / *min*
Lateral channels in flat terrain	1	10
Collector in flat terrain	2	5
Collector or lateral channel at over 2% terrain slopes	5	2
Collector or field channel at over 4% terrain slopes	10	1
Highway	10	2
Trunk road	5	5
Voivodeship road	2	10
Other road	1	

Channel retention time (t_r) is the time required for filling the irregularity in the channel until the presence of computed flow in the conduit. In accordance with this definition, the parameter is calculated from the following equation:

$$t_r = \frac{F \cdot L}{60 \cdot Q_{\mathrm{obl}}} \tag{3.7}$$

where: L – channel length (m), t_r – channel retention time (min), F – flow cross-sectional area (m^2) in the channel corresponding to the Q_{obl} flow in steady flow conditions.

Its size for the inner diameter of the channel (D) with the wall roughness coefficient (n) and longitudinal slope (J) of the circular conduit can be determined with the successive iteration method based on the following equation:

$$Q_{\mathrm{obl}} = F \cdot v_{\mathrm{sr}} = F \cdot C(h) \cdot \sqrt{R_h J} = \frac{F}{n} \cdot R_h^{2/3} \cdot J^{0.50} \tag{3.8}$$

For partially filled conduits, the size of the hydraulic radius is calculated with the following formula (see Figure 3.2):

$$R_h = \frac{D \cdot (\phi - \sin\phi)}{4 \cdot \phi} \tag{3.9}$$

In the threshold rainfall intensity (TRI) method, the calculation of the stormwater stream and pipe diameters starts from the upper sections of the sewer network (Figure 3.3). First, the velocity of the stormwater flow ($v_{\mathrm{zał}}$) in section 1 is specified, and the stormwater flow time through the channel (t_{p1}) is calculated using Eq. 3.1. In the next stage, the duration of design rainfall (t_{dm1}) is determined for the assumed rainfall frequency (C) and concentration time (t_k), assumed in accordance with Table 3.1.

Next, the calculated time of design rainfall is compared with the assumed rainfall duration, and the larger value is taken for further analysis. In the next stage, the rainfall intensity and the flow in channel 1 at the cross-section closing the modelled catchment are determined as follows:

$$Q_1 = q \cdot A_{\mathrm{zr1}} \tag{3.10}$$

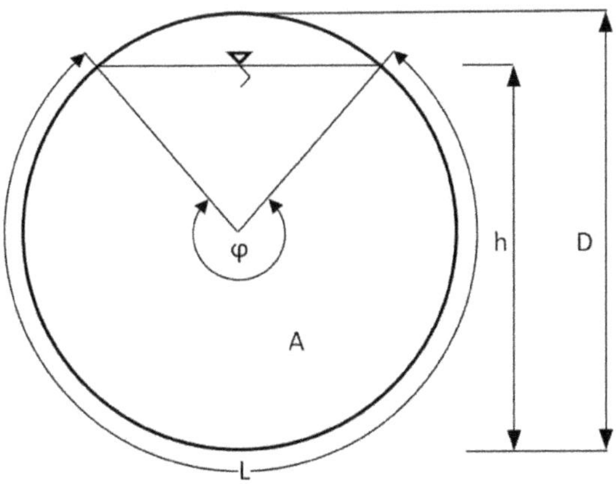

Figure 3.2 Calculation scheme of a circular conduit.

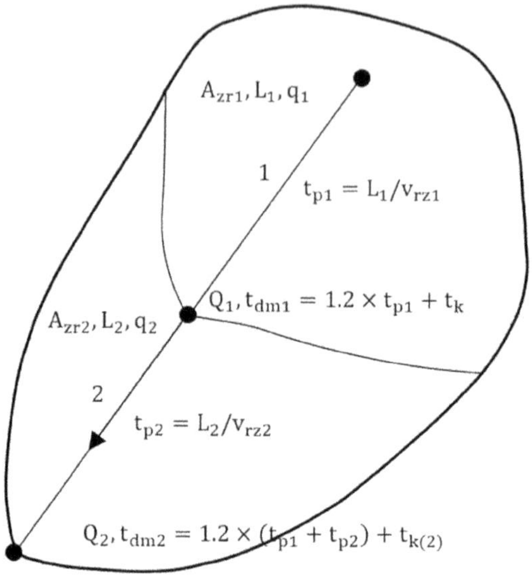

Figure 3.3 Catchment calculation scheme in the TRI approach (Kotowski 2011).

where: A_{zr1} – reduced area of partial catchment of channel 1 (ha) calculated using Eq. 3.3, q – rainfall intensity (dm³/ha·s).

On the basis of the Q_1 value, the conduit filling parameter $h=f(\phi)$ is determined for the assumed slope (J), roughness coefficient (n), and diameter (D). This is used for the determination of the actual flow velocity (v_{rz1}). By comparison of the assumed velocity ($v_{zał}$) with the actual velocity (v_{rz1}), the required accuracy of the calculation is checked; the difference between $v_{zał}$ and v_{rz1} should not exceed 0.1 m/s. When the calculation accuracy is insufficient, the procedure should be repeated for a corrected flow velocity. Once the accurate result meeting the condition of $v_{zał} - v_{rz1} < 0.1$ m/s is achieved for channel 1, calculations for channel 2 are carried out. The flow velocity ($v_{zał2}$) is assumed, and the sewer flow time (t_{p2}) is determined. Next, the design rainfall duration is calculated for the cross-section closing the catchment of channels 1 and 2 using the following formula:

$$t_d = t_{k(2)} + t_{p_1} + t_{p_2} + t_{r1} + t_{r2} \quad (3.11)$$

where: $t_{k(2)}$ – concentration time in channel 2 catchment (min), t_{r1}, t_{r2} – channel retention time in channel 1 and 2 (min), or:

$$t_d = t_{k(2)} + 1.2 \cdot \left(t_{p_1} + t_{p_2}\right) \quad (3.12)$$

In the next step, the rainfall intensity (q_2) and flow (Q_2) in the cross-section closing the channel 1 and 2 catchment are calculated using the formula:

$$Q_2 = q_2 \cdot \left(A_{zr1} + A_{zr2}\right) \quad (3.13)$$

For the calculated Q_2 value and the assumed conduit diameter D_2, the channel flow velocity (v_{rz2}) is determined in accordance with the recommendations described above and compared with the assumed velocity ($v_{zał2}$). When their difference is below 0.1 m/s, calculations for the channel network can be carried out as follows; otherwise, the procedure is repeated with a modified velocity value $v_{zał2.}$

3.3 Delay factor method (DFM)

The actual rain intensity observed in a given area (Imhoff & Imhoff 1996; Akan & Houghtalen 2003; Królikowska & Królikowski 2012) is used for calculation of the maximum outflow and the design rainfall duration; when these data are not available, they are determined from rainfall intensity maps or empirical formulas. In the DFM method, the outflow delay factor ($\phi_{t_d,C}$) as a function for rainfall duration td and rainfall frequency C is the basis for the calculations (Kotowski 2011):

$$\phi_{t_d,C} = \frac{38}{t_d + 9} \cdot \left(\sqrt[4]{C} - 0.3684\right) \tag{3.14}$$

In this method, the maximum stormwater stream is determined taking into account the terrain slope, surface imperviousness, and rainfall frequency, introducing the so-called peak flow runoff coefficient (ψ_s). The coefficient expresses the ratio of outflow (Q) to the product of rainfall intensity (q) and catchment area (A), and its values are available in the tables of guidelines ATV-A 118 (1999). In partial catchments, the values of the flow coefficient (ψ_s) are assumed constant. The minimum rainfall duration is determined based on the terrain category and catchment imperviousness (Table 3.2), which reflects the percentage area of roofs, roads, pavements, parking lots, yards, entrances, etc. in the entire catchment area.

The flow intensity in the cross-section closing the partial catchment is defined as:

$$Q = \phi_{t_d,C} \cdot \psi_s \cdot q \cdot A \tag{3.15}$$

where: $\phi_{t_d,C}$ – outflow delay coefficient characteristic for rainfall duration t_d and rainfall frequency C calculated from Eq. 3.1.

In the DFM method, the rainfall duration (t_d) is determined for each network calculation point (connection manholes). It is calculated with the assumption that the time of concentration (t_k) and channel retention time (t_r) $t_k = t_r = 0$. This indicates that the rainfall duration is equal to the flow time (t_p) along the length of channels located above the design point. Peak outflow coefficients and reduced areas of partial catchments are determined for the assumed rainfall frequency (C), catchment imperviousness, and terrain slopes.

Table 3.2 Minimum design rainfall duration specified by (Kotowski 2011)

Mean terrain slope (I_t)	*Imperviousness area*	*Minimum rainfall duration (t_{dm}), min*
<1%	<50%	15
	>50%	10
1%–4%		10
>4%	<50%	10
	>50%	5

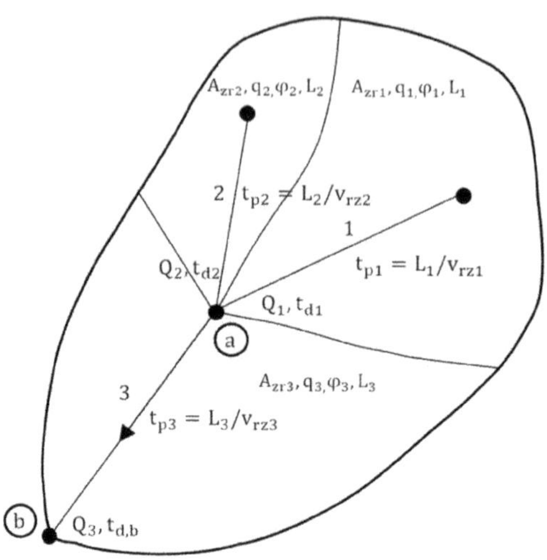

Figure 3.4 Catchment calculation scheme in the DFM approach (based on Kotowski 2011).

As in the TRI method, the calculations begin with consideration of manholes connected to channels draining the margins of the catchment (Figure 3.3). The calculation with the DFM approach is illustrated for a system of channels 1, 2, and 3 (Figure 3.4). First, the flow velocity in channel 1 ($v_{zał1}$) is assumed, and the flow time (t_{p1}), equal to the rainfall duration (t_{d1}) at modelled point "a", is determined. When it is shorter than the design rainfall duration, its minimum value is assumed (Table 3.2). For individual calculation points (including consecutive manholes), where this relationship is fulfilled, a constant outflow delay factor (ϕ_1), determined from the Eq. 3.14, is assumed. This is used for determination of the flow intensity (Q_1) from Eq. 3.1 in the cross-section closing the catchment of channel 1 and for calculation of flow velocity (v_{rz}) and duration (t_{prz}) along its length. When the difference between t_{prz} and t_{p1} is greater than 5%, the calculations are repeated for the modified flow velocity in channel 1. When satisfactory accuracy is achieved, calculations for channel 2 can be carried out (Figure 3.4). The procedure is the same as for channel and catchment 1, i.e. it consists in determination of flow duration (t_{p2}), rainfall duration (t_{d2}), and outflow delay factor (ϕ_2), which are the basis for calculation of flow intensity (Q_2) in the cross-section closing the catchment.

In design point "a" (Figure 3.4) at the junction of channels 1–2 (inlet to channel 3), the corrected flow (Q) is calculated, and its value depends on (Kotowski 2011):

$$Q_2 \le \frac{t_2}{9} \cdot Q_1 \tag{3.16}$$

where: Q_1 – flow in the channel with a shorter flow time, Q_2 – flow in the channel with a longer flow time, t_2 – longer flow time (min).

When this condition is fulfilled, the corrected flow (Q) is determined using the following formula:

$$Q = \phi_1 \cdot \left(Q_1 + \frac{t_1}{t_2} \cdot Q_2 \right) \tag{3.17}$$

where: ϕ_1 – delay coefficient for the channel with a shorter flow time, t_1 – shorter flow time (min).
Otherwise, the flow is determined using the following equation:

$$Q = \phi_2 \cdot \left(Q_1 + Q_2\right) \tag{3.18}$$

where: ϕ_2 – delay factor for a channel with a longer flow time.
In further analyses, attention should be paid to correct determination of design rainfall duration (t_{d3}) for design point "b". In this case, it is determined with the following equation:

$$t_{d,b} = t_1 + t_{p,3} \tag{3.19}$$

where: t_{p3} – flow time in the channel 3 section (min).
The other parameters are the outflow delay factor (ϕ_3) established with Eq. 3.14, and flow intensity in the cross-section closing the catchment of channels 1, 2, and 3 is determined as follows:

$$Q_3 = \phi_3 \cdot q \cdot \left(A_{zr1} + A_{zr2} + A_{zr3}\right) \tag{3.20}$$

where: q – rainfall intensity for rainfall duration ($t_{d,b}$) and frequency (C), dm³/ha·s; $A_{zr1,2,3}$ - reduced catchment area of channels 1, 2, and 3 calculated using Eq. 3.3 with the flow coefficient provided by guidelines ATV-A 118 (1999).

Further calculations are carried out following the principles described above.

The CRI, TRI, and DFM approaches presented above are typical methods for calculations and design of stormwater sewer systems. Due to their shortcomings and limitations, the rational method has been modified over the last 40 years. Interesting considerations were presented by Akan (1985) and Akan and Houghtalen (2003), who used the kinematic wave equation, taking into account the impact of terrain slope, roughness coefficient, and topography on the modelled maximum flow value. Guo (2000) introduced a modification that facilitated determination of catchment outflow hydrographs for any temporal rainfall intensity distributions. In turn, Jiapeng proposed formulas taking into account the catchment shape (Mazurkiewicz et al. 2020). In the cases considered by the authors, high agreement of calculation results with measurements was achieved. However, this method was not tested in other catchments; hence, this approach was not continued. To answer the need to reduce stormwater discharged into the sewer system, Guo & Urbonas (2009) proposed another modification of the rational method. It consisted in the transformation of the stormwater stream based on the observation that the stormwater flowing from an impervious surface does not enter the sewer system but a pervious area. This approach seems interesting and useful from the theoretical point of view regarding the analysed issue. However, the implementation of the approach is limited to its use in the case of single-family buildings only at the design stage. Akan and Houghtalen (2003) introduced a modification of the rational method by inclusion of the kinematic wave equation, but its application is limited to small homogeneous urban catchments with appropriate geometry and terrain with different characteristics (terrain imperviousness, roughness coefficient, terrain slope).

3.4 Hydrograph methods

Hydrograph calculation methods can be divided into hydrological and hydrodynamic approaches. In hydrological methods, the processes of time shift and retention on the catchment surface are usually calculated indirectly. Typically, the maximum flow value and the design rainfall duration are the results of calculations made with these methods. These approaches include the constant and limit rainfall intensity, the delay factor, and the flow duration factor (ATV-A 118 1999). A separate group of hydrograph calculation methods comprises hydrodynamic methods based on the physico-hydraulic patterns of channel flow described by the Saint Venant equation. These models include the impact of the types of land use in the catchment, delay in stormwater outflow resulting from hydrological processes in the catchment (catchment retention filling, infiltration, evapotranspiration), variability of rainfall intensity distribution in the catchment, directions of movement of rainfall fronts, and hydraulic phenomena in the conduit network (stormwater accumulation and flooding, backwater, etc.). These phenomena have been included in numerous programmes developed by commercial companies or research centres. The most common of these are SWMM, MOUSE, MIKE URBAN, etc.

3.5 Deterministic methods

The most recent hydrological models are deterministic models in which physical processes are the basis for the description of individual phenomena occurring during surface runoff. In these methods, the catchment is divided spatially into small areas for which the phenomena of interest can be considered homogeneous. Such a division is possible only with the use of digital maps. The methods take into account the spatial model, the distribution of atmospheric phenomena, and the characteristics of the catchment area with its terrain features and cover. Since not all processes have been identified and described, new models are being generated. The basis for calculation of the catchment inflow hydrograph in physical models is the Saint Venant differential equation system for conservation of mass and momentum. Richards & Bousinesq equations are employed for the description of groundwater movement (Clément et al. 2023). There are relatively many deterministic models for modelling non-urbanised catchments, e.g. CASC2D (Marsik & Waylen 2006), WASH123D (Shih & Yeh 2010), and SWAT (Zhang et al. 2014). The number of such models is substantially smaller in the case of urbanised catchments, since a model including large differences in land use, potential changes in land slopes, and changeable direction of stormwater flow along the surface is complicated and requires implementation of appropriate mathematical algorithms (Jacobson 2011; Jiang et al. 2018; Jato-Espino et al. 2019) and precise input data presented in digital maps. Additionally, in the event of stormwater overflow onto the ground, the mathematical description of the phenomenon becomes more complicated, as it is necessary to model the flow through the sewer network simultaneously with determination of the depth of the stormwater flooding layer and its flooding surface area, as well as the direction of the surface flow (Leandro et al. 2009; Obermayer et al. 2010; Djordjević et al. 2014). Simultaneous consideration of all these factors is complicated and is currently being investigated by many researchers (Brown et al. 2007; Leandro & Martins 2016). This may indicate that this issue will be solved and that it will be possible to model the stormwater flooding phenomenon fully consciously and reliably with consistent results. Nevertheless, due to the increasing complexity of the models, the number of coefficients to be calibrated, and data reliability, there is no easy and comprehensive solution to the problem, and this issue requires further research.

Chapter 4

Surface runoff generation and concentration

4.1 Surface runoff generation

Surface runoff is a complex of hydrological and hydraulic processes. Hydrological processes include runoff losses, and hydraulic processes denote the processes related to the movement (flow) of water along the surface of a catchment up to the point where it flows into the sewer network (runoff generation and runoff concentration). Thus, it can be concluded that surface runoff consists of two well-distinguishable phases, namely runoff generation, which is described by hydrological processes, and runoff concentration, which, on the other hand, is subject to the laws of hydraulics.

The process of runoff generation is influenced by a number of factors, which are denoted as runoff losses. These include interception, surface wetting, evaporation, evapotranspiration and, on pervious subsoil, infiltration. Surface runoff cannot occur until all runoff losses have been completed. During the rain, the individual losses do not increase linearly, nor do they begin or end at the same time. At the beginning of the rain, the basin must become wet (losses by surface wetting), the rain partly catches vegetation (losses by interception), and at the same time during the whole rain there is evaporation from soil and plants (evapotranspiration losses). Only after wetting does the water begin to infiltrate into the soil (infiltration losses). When the intensity of the rain exceeds the infiltration capacity of the soil, smaller or larger unevenness of the terrain surface begins to fill. If all the irregularities are filled in and the rain still lasts, surface runoff will start.

After the rain, the depressions are emptied by evaporation and infiltration, and the infiltration capacity in the soil is regenerated. Figure 4.1 schematically shows the time course of runoff formation.

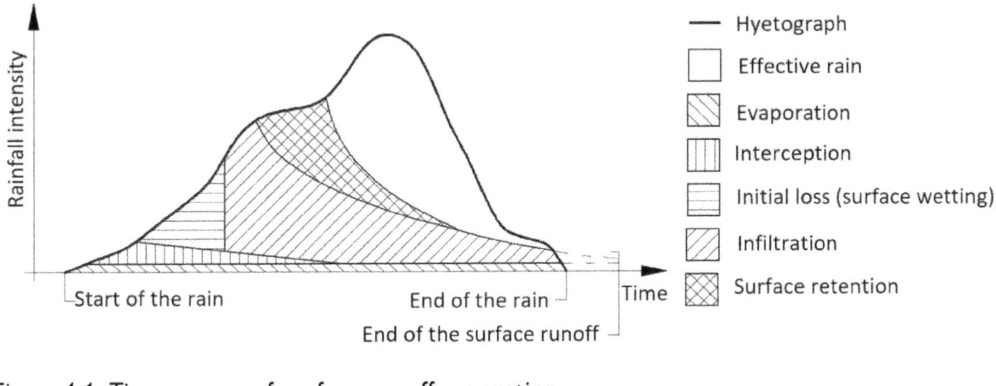

Figure 4.1 Time course of surface runoff generation.

DOI: 10.1201/9781003546436-4

Table 4.1 Parameters in the Green-Ampt model (Rossman 2009)

Type of soil	*$K(D_p)$, (mm/h)*	*D_p, (-)*	*P_f, (mm)*	*Porosity (–)*
Sand	120.39	0.346	49.02	0.437
Loamy sand	29.97	0.312	60.96	0.437
Sandy loam	10.92	0.246	109.98	0.453
Loam	3.30	0.193	88.90	0.463
Loamy silt	6.60	0.171	169.93	0.501
Sandy-clayey loam	1.52	0.143	219.96	0.398
Clayey loam	1.02	0.146	210.06	0.464
Silty-clayey loam	1.02	0.105	270.00	0.471
Sandy clay	0.51	0.091	240.03	0.430
Silty clay	0.51	0.092	290.07	0.479
Clay	0.25	0.079	320.04	0.475

Note: θ – initial soil water deficit.

Knowledge of runoff losses is an initial condition for runoff modelling. Incorrect loss entry can lead to inaccurate calculations of effective rainfall and, consequently, runoff generation (e.g. concentration time or runoff volume from the basin).

The infiltration rate is described in the Green Ampt model as (Akan & Houghtalen 2003):

$$f_{(t)} = \frac{K(D_p)\cdot(Z + P_f)}{Z} \tag{4.1}$$

where: $K(D_p)$ – hydraulic water permeability (mm) (Table 4.1), P_f – capillary suction height (mm) (Table 4.1), Z – thickness of the saturation zone measured from the ground surface expressed in mm, D_p – initial soil water deficit.

An alternative approach to the aforementioned methods for determination of infiltration is the SCS method (Hawkins 1993; Akan & Houghtalen 2003), which is commonly used to determine the effective rainfall in agricultural catchments. In contrast to the other methods described above, in the approach developed by the US Soil Conservation Service, the variability of the infiltration value during the rainfall event is not modelled. The method is based on the assumption that the effective rainfall depends on the type of soils in the catchment, the mode of land use, and the soil moisture at the beginning of the calculation period. The SCS method assumes that these factors and their variability over time are captured in the CN value (Ignar et al. 1995; Tedela et al. 2012). However, in the case of urban catchments, it is often problematic to determine the type of soil; moreover, the infiltration value fluctuates largely during the investigated rainfall events. Thus, the CN values determined based on literature data (Schneiderman et al. 2007; Rutkowska et al. 2014) and the calculated values may vary significantly. This may generate problems with calibration of the model, and no satisfactory agreement between the calculation and measurement results can be achieved.

4.1.1 *Initial loss (micro retention, wetting)*

The portion of rainfall that does not enter the channel due to interception, infiltration, and replenishment of soil moisture is retained in depressions (Table 4.2). It is dependent on the potential

Table 4.2 Values of μ for the calculations of initial loss S_p

CN (curve number)	μ
CN < 70	0.075
70 ≤ CN < 80	0.100
80 ≤ CN < 90	0.150
90 ≤ CN	0.200

catchment retention (maximum retention value after a very long rainfall), as the following are part thereof:

$$S_p = S \cdot \mu \tag{4.2}$$

where: S – potential catchment retention, μ – empirical coefficient according to the Table 4.2.

Impervious area retention, i.e. wetting of the cover, is the process of forming a thin layer of water that either starts the wetting process or is necessary for run-off to form on the surface. It consists in wetting the areas where no infiltration process occurs. The impervious area retention depends on the land cover, land use, and development of the catchment. The extent and types of impervious areas can be determined using aerial photographs, large-scale plans, and topographic maps. Wetting losses are as follows: impervious areas (asphalt, roofs, paved roads, gravel roads) – 0.20–0.50 mm; pervious surfaces (gardens, parks, arable land) – 0.20–2.00 mm. Water retained in depressions is within the following ranges: very smooth impervious areas (0.20–0.40 mm), smooth impervious areas (0.50–0.70 mm), areas covered with sparse vegetation (0.60–2.50 mm), areas covered with dense vegetation (2.50–4.00 mm).

Surface soil layer retention (soil detention) is the amount of water for wetting the near-surface soil layer. Open water retention is considered as the capacity of water at the initial time. Open channel retention is formed from the rainfall falling on the water surface of channels, watercourses, trenches, and canals. Its amount depends on the density of the watercourse network, and its length and width.

4.1.2 *Interception*

The process of periodic retention of rainfall by a plant community is called interception. It constitutes the portion of rainfall that evaporates into the atmosphere after wetting the plant surface. Of all the processes that determine wetting and rainfall loss on a catchment surface, the interception pattern of the plant cover is the most well-known. The starting point is a homogeneous patch of natural or cultivated vegetation. Leaf geometry is of great importance, and their position at different heights affects the rate of dripping.

4.1.3 *Evapotranspiration*

Actual evapotranspiration is the water vapour released into the atmosphere as a result of evaporation from the soil and transpiration of vegetation under existing meteorological conditions and the current state of soil moisture. Potential evapotranspiration is defined as the maximum amount that could be released into the atmosphere as water vapour under given meteorological conditions from soil completely covered with vegetation and saturated with water.

Actual evapotranspiration, i.e. evaporation from the soil surface (evaporation) and from plants (transpiration), is the basic component of the water balance. It depends on meteorological factors (solar radiation, astronomical day length, sunshine, air temperature, air humidity deficit, wind speed), hay yields and the associated size of the evaporating leaf area, as well as soil moisture. The accuracy of estimating actual evapotranspiration at the grassland water needs balancing stage significantly determines the accuracy of calculating the other components of the water balance equation during decadal and regrowth periods.

4.1.4 *Infiltration*

Infiltration is the process of water entering the soil. It constitutes a significant loss on pervious and semi-pervious surfaces. Its values are an order of magnitude greater than losses from interception, surface wetting, and evaporation on pervious surfaces.

In the hydrology of urbanised areas, there are two approaches to calculating infiltration losses. The first method is based on the physical properties of the soil, such as hydraulic conductivity and capillary suction. This approach is most prominently represented by the Green and Ampt equation (Green & Ampt 1911). The disadvantage of this method is the requirement for detailed knowledge of the soil's physical characteristics, which, in most cases, are not available. Consequently, the entire calculation often relies on the estimation of several parameters, usually based on data published in the literature.

The second method is the hydrological approach, which is essentially a parametric method that uses lumped soil parameters to simulate infiltration. These parameters are often optimised and adapted during infiltration calculations to achieve results consistent with laboratory or field measurements. The best-known equation in this category is the Horton equation (Horton 1939; 1940), in which the key parameter is the coefficient describing the decline of infiltration over time.

The instantaneous infiltration rate v_i and the cumulative infiltration f are in this case expressed by:

$$v_i = v_c + \left(v_0 - v_c\right) \cdot e^{-kt} \tag{4.3}$$

$$f = v_c \cdot t + \frac{1}{k}\left(v_0 - v_c\right) \cdot \left[1 - e^{-kt}\right] \tag{4.4}$$

where: v_i – the instantaneous infiltration rate (m/h), v_0 is the initial (maximum) infiltration rate (m/h), v_c – steady infiltration rate (equal to saturated hydraulic conductivity) (m/h), f – cumulative infiltration (m), k – decrease coefficient (h^{-1}) and t is the time since onset of infiltration (h).

The value of the decrease coefficient k varies from 0.67 h^{-1} up to 49 h^{-1}, but most values are in the range of 3–6 h^{-1} (0.00083–00167 sec^{-1}). Lower values apply to sandy soils. Field measurements of Huber et al. (Huber et al. 1981) and Geiger et al. (Geiger et al. 1987) recommend using 4.14 h^{-1} (0.00115 sec^{-1}), which is a predetermined value e.g. also in SWMM (Table 4.3). Using this value implies that the infiltration rate decreases by 98% compared to its initial value within the first hour of infiltration. The influence of the decrease coefficient k on the infiltration rate is shown in Figure 4.2.

Table 4.3 Average values of initial (v_i), steady infiltration rate (v_c) and decrease coefficient to Horton equation (Chow 1964; Huber et al. 1981)

Soil type	v_i *(cm/h)*	v_c *(cm/h)*	*k* *(h⁻¹)*
Sand, sandy soil	8.0[a]–20.0[b]	1.0–1.6	3
Clay sand	3.0[a]–8.0[b]	0.5–1.0	4
Sandy clay, clay	3.0[a]–8.0[b]	0.2–0.5	5
Clay, loamy soil	1.5[a]–3.0[b]	0.0–0.2	6

[a] In the dry state without vegetation,
[b] In the wet state with dense vegetation.

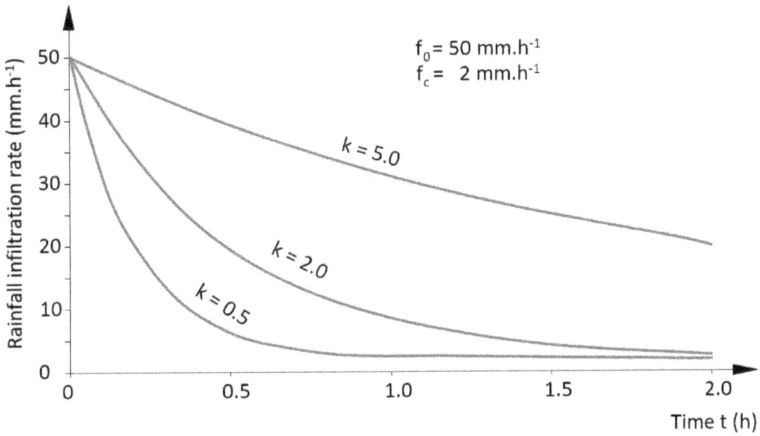

Figure 4.2 Influence of the decrease coefficient *k* on the infiltration rate.

The equations above describe infiltration into the soil during rain. After the rainy season, the rate of soil infiltration regenerates depending on many factors. In simple terms, it can be expressed by the following equation:

$$v_i = v_0 + \left(v_T - v_0\right)\exp(-\Gamma \cdot T) \tag{4.5}$$

where: v_i – the instantaneous infiltration rate (m/h), v_0 – the initial (maximum) infiltration rate (m/h), T – the time since the end of the rain (h), Γ – the regeneration constant (h^{-1}).

The regeneration rate of the infiltration rate is shown in Figure 4.3.

4.1.5 Surface retention

Stormwater on the ground surface, which wets various types of cover, does not always infiltrate into the soil. Instead, it can accumulate on slopes or valley terraces, where it is retained temporarily or permanently as surface retention. This process is closely linked to the terrain's relief, undrained depressions, watercourses, lakes, and similar features. During the initial phase of rainfall, the catchment's initial surface retention state plays a key role. Together with rainfall intensity, this factor influences the development of soil retention and, ultimately, the subsurface zone.

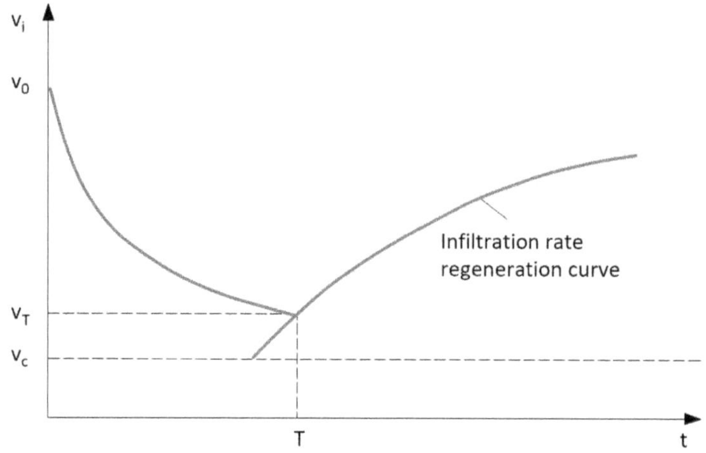

Figure 4.3 Regeneration of infiltration rate in the dry period.

Table 4.4 Surface retention for different types of surfaces (Akan & Houghtalen 2003)

Surface	*Surface retention capacity values h_k (mm)*
Very smooth impervious, no vegetation	0.4–0.6
Smooth impervious, no vegetation	0.7–0.9
Bare soil, grassy surface, sparse vegetation	0.8–2.7
Surface with dense vegetation	2.7–4.2

Surface retention, which inhibits surface runoff, may then develop either through infiltration or storage within slopes and valleys. In a catchment, several types of retention can be distinguished: impervious areas, the surface soil layer, channels and open water, and depressions.

Surface retention is one of the most significant runoff losses. On impervious surfaces without infiltration, it begins to appear immediately after the surface becomes wet, and on semi-pervious and pervious surfaces together with infiltration. On pervious surfaces, retention occurs when the intensity of rain is higher than the unsaturated hydraulic conductivity of the soil, and water begins to accumulate in the depressions and unevenness of the soil. Irregularities can range in area size from a few square centimetres to a few square metres. This is the volume that must be filled with water before surface runoff from the depression areas appears. Table 4.4 shows the surface retention values for different surface types and different gradients.

These values also include surface wetting losses, which according to Pecher (1969) should be considered as 0.2 to 0.5 mm for impervious surfaces and 0.2–2.0 mm for pervious surfaces. Kidd (1978) recommends for residential areas an average surface retention capacity of 1.5 mm. When determining the effective rain for pervious areas, surface retention is applied above the infiltration curve.

4.1.6 *Permanent losses*

Permanent losses represent the rainfall that never reaches the sewer network due to partially pervious surface structure (also asphalt), evaporation during rain, clogged drains, or incorrect

Table 4.5 Retention values for different types of surfaces (Chow et al. 1988)

Type of surface	*Retention value* $d_{imp(per)}$ *(mm)*
Impervious surfaces	1.25 ÷ 2.50
Lawns	2.54 ÷ 5.08
Pastures, meadows	5.08
Forest litter	7.62

slope conditions. Some of the rain from impervious areas can reach lawns, etc. A certain part of the surface runoff in areas with individual development with fenced plots does not infiltrate into the network but infiltrates in the front gardens. Furthermore, in the autumn, part of the rain is caught on the leaves collected on the roads along the kerbs.

The value of permanent losses can be determined approximately depending on the maximum value of the volume runoff coefficient $\psi_{0,\max}$

$$t_s = 1 - \psi_{0,\max} \tag{4.6}$$

Permanent losses can play an important role, especially in calibrating runoff concentration time and total effective rain. In some models, e.g. in the MOUSE model (DHI 1998) to express the percentage of permanent losses, the so-called hydrological reduction factor is introduced. This means that the runoff coefficients from the individual areas are multiplied by this factor, which expresses the permanent losses.

4.2 Surface runoff concentration

4.2.1 *Isochron method*

The idea of the isochron method is based on the assumption that the run-off hydrograph is formed by the concentration of rainwater run-off from rainfall of constant intensity. Assuming a certain run-off interval ($\Delta\tau$), lines of simultaneous run-off of a certain time length can be determined in the catchment (Figure 4.4).

Thus, a catchment of area F can be divided into a number of elementary plots depending on the run-off time, where the flow time from the most distant point of the catchment to the closing

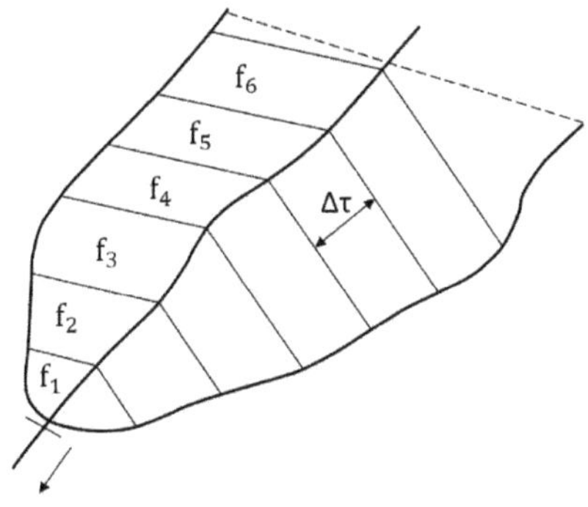

Figure 4.4 Computational diagram of the catchment area.

cross-section is equal to $\tau = \tau_{max}$. The amount of run-off at time $\tau_1 = \Delta\tau$ corresponds to the run-off from the first plot and equals:

$$Q_1 = f_1 \cdot s_1 \tag{4.7}$$

where: s_1 – the depth of the water layer that the catchment releases in the first time unit, in turn, the run-off in the second time unit is:

$$Q_2 = f_1 \cdot s_1 + f_2 \cdot fs_2 \tag{4.8}$$

where: s_2 – depth of the given layer of water in the second unit of time, which can generally be written as:

$$Q_t = \sum_{\tau=1}^{\tau=t} f_\tau \cdot h_{t-\tau+1} \tag{4.9}$$

Or in integral form:

$$Q(t) = \int_{\tau=0}^{\tau=t} s_{t-\tau} \cdot b_\tau \cdot v_\tau \, d\tau = \int_{\tau=0}^{\tau=t} s_{t-\tau} \, dF_\tau \tag{4.10}$$

where: b_τ – plot width, v_τ – run-off velocity, $S_{t-\tau}$ – depth of the water layer outflowing from separated area per unit time τ is referred to the impervious surface in engineering practice and is determined using the formula (Akan & Houghtalen 2003; Banasik 2009).

Equation 4.10 presented above describes the variability of catchment run-off as a function of run-off velocity, catchment width, and catchment area. However, due to limited calculation capabilities at the method development stage, the modelling of the variability of the afore-mentioned variables in the equation was limited. Therefore, the following simplifications were adopted: constant catchment width, constant surface runoff intensity, i.e.:

$$h_1 = h_2 = h_3 = \cdots = h_t \tag{4.11}$$

Using the isochron method, surface run-off hydrographs (Figures 4.5–4.7) were determined for the following cases:

a $\tau_{max} < T$
b $\tau_{max} > T$
c $\tau_{max} = T$

where: T – rainfall duration.

In the first case, the time of runoff is longer than the run-up time (time required for the hydrograph/flow to form to a certain cross-section, i.e. the profile that closes the catchment). Initially, the runoff from the catchment will increase, since there will be a steady increase in the number of plots participating in the runoff. This will continue until the whole catchment takes part in the runoff, i.e. until time τmax, which can be written as:

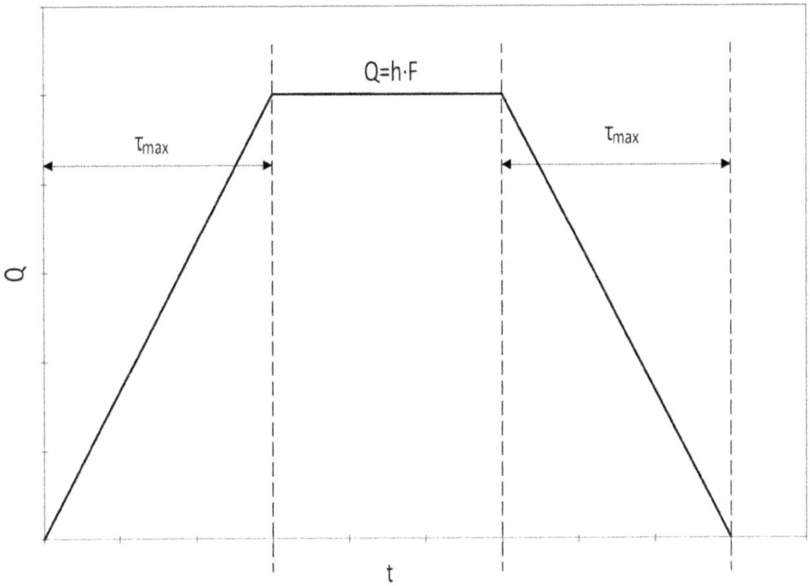

Figure 4.5 Run-off hydrograph for $\tau_{max} < T$.

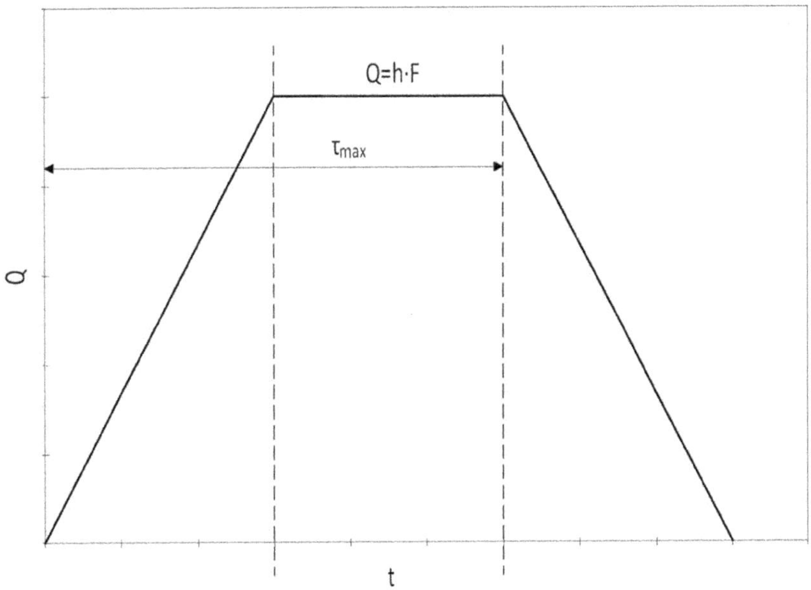

Figure 4.6 Run-off hydrograph for $\tau_{max} > T$.

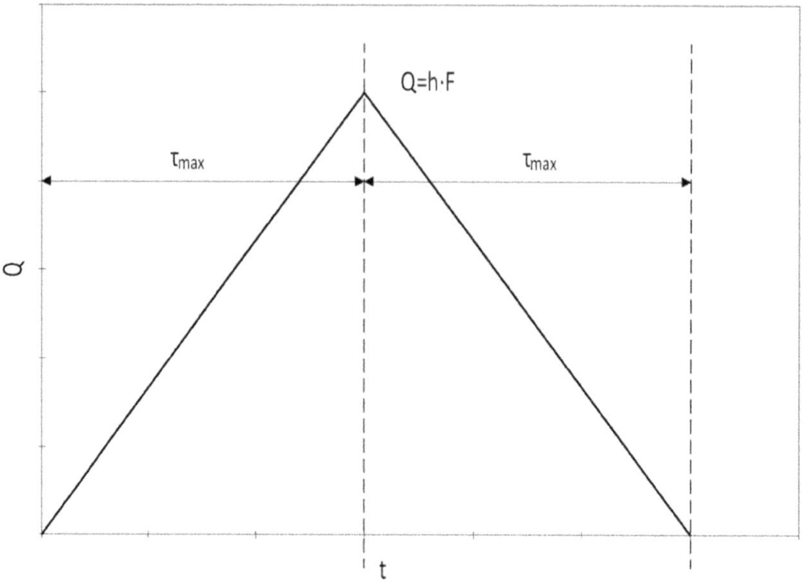

Figure 4.7 Run-off hydrograph for $\tau_{max} = T$

$$Q_1 = h \cdot f_1$$

$$Q_6 = h \sum_{1}^{6} f = h \cdot F \tag{4.12}$$

or in general

$$Q(t) = h \cdot \sum_{\tau=1}^{\tau=t} f_\tau \tag{4.13}$$

For a time period $\tau_{max} \leq t \leq T$, the water will runoff from the entire catchment simultaneously, which means that:

$$Q(t) = h \cdot \sum_{\tau=1}^{\tau=\tau_{max}} f = h \cdot F \tag{4.14}$$

will be a horizontal line. For $T \leq t \leq \tau_{max} + T$ the size of the catchment area from which the water is drained will decrease linearly, giving the relationship:

$$Q(t) = h \cdot \sum_{\tau=t-1}^{\tau-\tau_{max}} f_\tau \quad (4.15)$$

and the obtained hydrograph will be a trapezoid (Figure 4.5).

When performing identical considerations for case (b), it can be concluded that the hydrograph plot will also be a trapezoid (Figure 4.6). The maximum runoff in the closing profile will occur due to water runoff from a part of the catchment, the magnitude of which will correspond to the runtime $\tau = T$. This runoff will persist until $t = \tau_{max}$, only the position of the FT area that forms it will change. The less the values of T and τmax differ from each other, the more the trapezoidal graph will resemble a triangle.

4.2.2 Linear reservoir method

In both urban and rural catchments, the occurring hydrological processes have a similar origin, since they result from the catchment response to rainfall. However, due to the specific nature of urban catchments, the large spatial variation of impervious areas in a relatively small area, the short land concentration, and the time of sewage flow through the channels, the dynamics of the occurring phenomena are faster than in rural catchments. This is due to the overlapping of stormwater runoff hydrographs of individual channels in the stormwater network, which can lead to backflow and stormwater flooding the land surface. In these analyses, the selection of a computational model for calculating runoff from individual catchments in an urbanised area is crucial. The volume of water changing in a catchment can be written generally with the following equation:

$$\frac{dS}{dt} = i - q \quad (4.16)$$

where: i – rainfall depth in the catchment, q – run-off from the catchment, S – volume of water in the catchment, dt – time step.

Assuming that the catchment run-off is a linear function $S(t)$:

$$q = a \cdot S(t) \quad (4.17)$$

The volume of water in catchment $S(t_2)$ after time t_2 is described with the following equation:

$$S(t_2) = \frac{S_0 + i \cdot \exp(a \cdot t)}{\exp(a \cdot t)} \quad (4.18)$$

where: S_0 – initial retention of the catchment, a – retention constant, t – time.

When the run-off from the reservoir is nonlinear and described by the $q = a \cdot S(t)^n$ (where: $n = 2$) the following relation occurs:

$$t_2 = \frac{\tanh^{-1}\left(\sqrt{\frac{a}{i}} \cdot S_2\right) - \tanh^{-1}\left(\sqrt{\frac{a}{i}} \cdot S_0\right)}{\sqrt{a \cdot i}} \quad (4.19)$$

For any value of n, equation (1) needs to be solved numerically due to the lack of analytical form of the S_2 calculation results.

4.3 Unit hydrograph

4.3.1 Model Nash

Literature data (Bhunya et al. 2007; Wang & Bi 2020) show that one of the first methods used to calculate catchment runoff was the unit hydrograph method. The concept of the unit hydrograph method was developed as early as 1932 by Sherman and further described by Chow (1964) and Banasik (2009). In this method, it was assumed that each catchment has a characteristic runoff hydrograph and the hydrograph bases resulting from rainfall of the same duration Δt are identical and do not depend on the amount of the so-called effective rainfall. The theoretical basis of the method assumes that the physical and geographic characteristics of the catchment (shape, area, land use, slope, type of soils, moisture content, etc.) affecting the runoff for each catchment are constant and dependent on local conditions. In this approach, a unit hydrograph is a plot of the course of a rainfall–runoff event constituting the catchment response, induced by an effective rainfall of unit depth and duration t. For unit rainfall $\Delta t \to 0$, it is called the instantaneous unit hydrograph. Out of a large number of rainfall–runoff models that make it possible to calculate instantaneous hydrographs and then hydrographs of catchment runoffs, taking into account the way of calibrating the coefficients included in them, the conceptual model described by Banasik (2009) turned out to be of particular interest for implementation in urban catchments. In this model, runoff from a catchment is compared to the flow through virtual reservoirs, and the stormwater captured in the reservoirs depends on the catchment retention and effective rainfall. The relationship $q = u(t)$ corresponding to the run-off from the last N-th reservoir is treated as an instantaneous unit hydrograph in the cross-section enclosing the catchment and is expressed as follows:

$$u(t) = \frac{1}{k \cdot \Gamma(N)} \cdot \left(\frac{t}{k}\right)^{N-1} \cdot \exp\left(-\frac{t}{k}\right) \tag{4.20}$$

where: N – number of reservoirs in the cascade, k – reservoir retention parameter, $u(t)$ – ordinates of the instantaneous unit hydrograph, $\Gamma(N)$ – gamma function, the value of which depends on the number of reservoirs in the cascade.

Unit hydrograph is described by the following parameters:

- time of maximum flow occurrence:

$$t_p = (N-1) \cdot k \tag{4.21}$$

- lag time:

$$\mathrm{LAG} = N \cdot k \tag{4.22}$$

- maximum $u(t)$ value:

$$u_p = \frac{1}{k \cdot \Gamma(N)} \cdot \frac{(N-1)^{N-1}}{\exp(N-1)} \tag{4.23}$$

In the case of rainfall and flow measurements, the values of the parameters N, k should be optimised to obtain the greatest agreement between the calculation results and the measurements. Otherwise, the empirical relationships given by Banasik (2009) determined for catchments smaller than 50 km^2, can be used for this purpose. The lag time is determined from the formula:

$$\mathrm{LAG} = 1.28 \cdot A^{0.46} \cdot \left(1 + \mathrm{Imp}\right)^{-1.66} \cdot P_{\mathrm{ef}}^{-0.27} \cdot t_{r,\mathrm{ef}}^{0.37}$$

$$k = 0.56 \cdot A^{0.39} \cdot \left(1 + \mathrm{Imp}\right)^{-0.62} \cdot P_{\mathrm{ef}}^{-0.11} \cdot t_{r,\mathrm{ef}}^{0.22} \tag{4.24}$$

where: $t_{r,ef}$ – effective rainfall depth (min), A – catchment area (km^2), Imp – catchment imperviousness, P_{ef} – effective rainfall depth is the remaining part of rainfall after taking into account the losses (wetting of the terrain, inception, infiltration, evaporation, filling of depressions). One of the most widely used methods for calculating effective rainfall is that developed by the Soil Conservation Service (SCS) in the USA. The effective rainfall depends on the type of soil, the catchment land use, and the moisture of soils in the catchment before the occurrence of the rainfall under study. All these factors are captured by the dimensionless parameter CN, taking values from 0 to 100. This parameter is functionally related to the maximum potential catchment retention S, expressed in mm

$$S = 25.4 \cdot \left(\frac{1{,}000}{\mathrm{CN}} - 10 \right) \tag{4.25}$$

The effective rainfall after time $t_j = j - \Delta t$ is calculated from the following formula:

$$H_j = \begin{cases} 0 & \text{when } P_j - 0.2 \cdot S \le 0 \\ \dfrac{\left(P_j - 0.2 \cdot S\right)^2}{P_j + 0.8 \cdot S} & \text{when } P_j - 0.2 \cdot S > 0 \end{cases} \tag{4.26}$$

where: S – maximum potential catchment retention; CN – method parameter; H_j – effective rainfall at time t_0 to t_j ($H_j = \sum_{l=1}^{j} \Delta H_l$); P_j – effective rainfall at time t_0 to t_j ($P_j = \sum_{l=1}^{j} \Delta P_l$).

In order to determine the total run-off hydrograph caused by successive effective rainfall of duration Δt each, the superposition principle is applied by summing the ordinates of the elementary unit hydrographs; the ordinates in each are calculated using the formula:

$$h_i = \frac{A}{3.6 \cdot \Delta t} \cdot \int_{t-\Delta t}^{t} u(\tau)\, d\tau \tag{4.27}$$

The total hydrograph is the resultant of unit hydrographs, caused by successive effective rainfall (for n – rainfall intervals), where the assumed m – ordinates are determined from the formula

$$Q_i = \sum_{j=1}^{i,n} h_k \cdot \Delta H_j \tag{4.28}$$

where: $k = i - j + 1$; $i = 1, 2, \ldots, m + n - 1$.

4.3.2 Gamma function (SCS)

In the SCS method, the runoff unit hydrograph can be presented as a triangular or curvilinear one, in which the maximum flow is expressed by the equation:

$$Q_m = \frac{\alpha \cdot A}{t_p} \tag{4.29}$$

where: α – an empirical coefficient depending on the characteristics of the catchment area; A – catchment area (km^2), t_p – time after which the maximum flow is determined as:

$$t_p = \frac{t_R}{2} + 0.6 \cdot t_c \tag{4.30}$$

where: t_R – rainfall duration, t_c – time of field concentration.

The curvilinear unit outflow hydrograph in the SCS method describes the relationship:

$$Q(t) = Q_m \cdot \left[\frac{t}{t_p} \cdot \exp\left(1 - \frac{t}{t_p}\right) \right]^{n-1} \tag{4.31}$$

where: n – shape parameter.

4.4 Time of concentration

Time of concentration determines how long it will take for a water droplet to move from the hydraulically furthest point in the catchment to the design profile (the section that closes the catchment). Time of concentration represents the hydrologic response time of an urban watershed. It is an important parameter used in many desktop hydrologic methods, such as the rational method. It is also a required watershed characteristic in more comprehensive rainfall–runoff models. The time of concentration of a catchment is calculated using the longest flow path within the catchment. It is sometimes referred to as the longest flow path or the hydraulic length of a watershed. The longest flow path is not always obvious. Frequently, there is a need to calculate the flow time along several paths before the one with the largest travel time can be identified as the longest path. Different flow types, such as overland and channel flows, can occur along a flow path. When this occurs, the travel times for the different flow segments have to be determined separately and summed up to determine the time of concentration.

The formula proposed by Akan and Houghtalen (2003) can be used to calculate the concentration time:

$$t_c = \frac{0.029 \cdot (n \cdot L)^{0.8}}{P^{0.5} \cdot J_0^{0.4}} \tag{4.32}$$

where: n – Manning's roughness coefficient, L – runoff length, P – rainfall depth, J_0 – terrain slope.

In the SCS method, the formula suggested for the calculation of concentration time for the values of $L < 190$ m is as follows

$$t_c = \frac{L}{3{,}600 \cdot \upsilon} \tag{4.33}$$

where: υ – mean flow velocity (m/s) determined from the Chezy–Manning's equation, which under appropriate assumptions can be written as:

$$\upsilon = \omega \cdot J_0^{0.50} \tag{4.34}$$

where: ω – values to 16 and 20.3 for impervious and pervious areas, respectively.

Akan and Houghtalen (2003), based on the kinematic wave theory, provided an equation to calculate the concentration time:

$$t_c = \frac{\left(L \cdot n\right)^{0.6}}{J_0^{0.3} \cdot i^{0.4}} \tag{4.35}$$

where: i – rate of rainfall excess. J_0 – average longitudinal slope of the terrain.

Kirpich's formula in the form presented below can be used to calculate the time of concentration (terrain, channel):

$$t_c = \frac{0.0195 \cdot L^{0.77}}{J_0^{0.385}} \tag{4.36}$$

For overland flow on concrete and asphalt surfaces, t_c is multiplied by 0.4 and concrete channels by 0.20. For overland flow on grass surfaces, it is multiplied by 2.0.

4.4.1 Kinematic wave

4.4.1.1 Kinematic wave application in SWMM

One of the most frequently used models in the SWMM program for simulation of outflow from partial catchments is the so-called nonlinear flat reservoir algorithm (Rossman et al. 2016; Leandro & Martins 2016). It is the basis for the transformation of rainfall into effective flow. In this model, surface runoff in separate catchments is calculated based on a system of differential equations of the stormwater volume for impervious and pervious surfaces:

$$\frac{dh}{dt} = I_{(t)} - E_{(t)} - \frac{1}{n_{imp}} W \cdot J^{1/2} \cdot \left(h - d_{\text{imp}}\right)^{5/3} \tag{4.37}$$

$$\frac{dh}{dt} = I_{(t)} - E_{(t)} - f_{(t)} - \frac{1}{n_{\text{imp}}} W \cdot J^{1/2} \cdot \left(h - d_{\text{perv}}\right)^{5/3} \tag{4.38}$$

Table 4.6 Manning's roughness coefficient for different surfaces (Chow et al. 1988; Linsley et al. 1958)

Type of surface	*Roughness coefficient n* ($m^{-1/3}\cdot s$)
Asphalt	0.015
Concrete	0.010 ÷ 0.015
Thick clay layer, compact loess, compact fine gravel	0.018 ÷ 0.019
Paving slabs	0.020 ÷ 0.022
Surfaces covered with rough crushed stone, field stone pavement	0.024 ÷ 0.026
Compact clay	0.030
Low grass	0.035
Tall grass	0.050
Ripe cereals	0.060
Arable soils	0.130
Vast area covered with shrubs and dense weeds	0.070
Sparse trees and shrubs	0.080
Various crops, meadows, stoneless soil	0.090 ÷ 0.110
Clear area with clumps of trees without outgrowths	0.050
Clear area with clumps of trees with numerous outgrowths	0.080
Sparse lawn	0.300
Moderate lawn	0.400
Dense lawn	0.800

where: J – mean terrain slope in the catchment, h – water layer depth on the catchment surface, $I_{(t)}$ – rainfall depth determined with regional rainfall models or based on rainfall measurement results, $E_{(t)}$ – evapotranspiration value based on data in the Hydrological Atlas, n_{imp}, n_{perv} – roughness coefficient for impervious and pervious surfaces based on literature data (Table 4.6) or determined at the model calibration stage, d_{imp}, d_{perv} – retention depth for impervious and pervious surfaces (Table 4.5), W – flow path width, $f(t)$ – infiltration function.

The flow path width is one of the key parameters determined during model calibration (Szeląg et al. 2016; Kiczko et al. 2018; Wang 1991; Krebs et al. 2013). There are many indirect methods for the determination of the flow path width, but their universality raises doubts, and the value of this parameter depends on catchment characteristics, partial catchment size, terrain slope, impervious to pervious surfaces, types of land use within the catchment, etc. As shown by the literature review (Sun et al. 2014; Ballinas-González et al. 2016; Rossman & Huber 2016), the flow path width in the catchment is often determined with the following formula:

$$W = \alpha \cdot \sqrt{A} \tag{4.39}$$

$$W = \frac{\beta \cdot A}{L_{kol}} \tag{4.40}$$

where: W – flow path width, L_{kol} – length of the main channel in the partial catchment, A – catchment surface area, α, β – empirical coefficients determined via calibration.

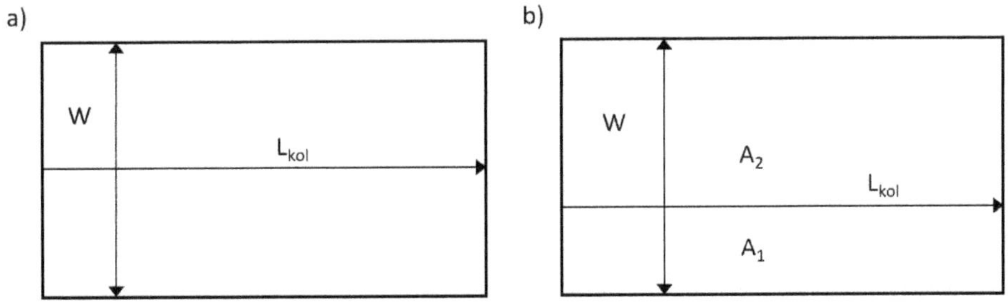

Figure 4.8 Scheme of a catchment with (a) symmetric, (b) asymmetric stormwater outflow channel

When the stormwater outflow channel is located symmetrically or asymmetrically in the partial catchment (Figure 4.8), the flow path widths in the considered catchments can be determined with the following formulas, respectively (Rossman 2009):

$$W = 2 \cdot L_{\mathrm{kol}} \tag{4.41}$$

$$W = \left(2 - S_k\right) \cdot L_{\mathrm{kol}} \tag{4.42}$$

where: L_{kol} – length of the collector discharging stormwater from the partial catchment (m), S_k – skewness index determined as follows: $S_k = |A_2 - A_1| \cdot |A_2 + A_1|$ (–).

Chapter 5

Flow in the sewer network

5.1 Fundamentals of hydraulic modelling of sewer networks

The considerations presented in this chapter are conducted in relation to water treated as the incompressible liquid dispersion phase of a multi-phase mixture such as sewage. In sanitary engineering, issues related to hydrodynamics appear, among others, in relation to hydraulic calculations of canals and conduits, understood as the determination of their shapes, dimensions, slopes, filling with sewage, and sewage flow velocities. In practice, two basic instances of performing hydraulic calculations for sewer conduits are encountered. The first and most frequent is designing sewer networks, where calculations are made for individual sections between specific trunk sewers, based on predetermined credible data on the flow rate, the adopted slope of the conduit, and the roughness of its walls (Kotowski 2011). The second case, when calculations of this kind are performed, relates to validation calculations and simulations of the operation of a sewer system designed or built. The inputs available in such a case include information concerning the shape and dimensions of the conduit and its filling, based on which the volumetric flow rate, the flow velocities in the conduit, and the stresses at the boundary between the sewage and the conduit walls are calculated.

The water flow in the sewerage network is a complex of hydraulic processes, which form the second component of numerical rainfall–runoff models (the first component is the calculation of surface runoff). In this process, the outflow hydrographs, defined outflow conditions (boundary conditions), and the initial state in the sewer network (initial condition) are considered as input values. The output of the runoff simulation is then the runoff hydrographs, water level heights, CSO volumes, etc.

In this chapter, the basic principles will be explained, and the control equations of pressureless as well as pressurised wastewater flow in sewers will be described.

In general, fluid flow can be divided into:

- unsteady (non-stationary), in which the hydraulic quantities depend on the position of the point and on time,
- steady (stationary), in which the hydraulic quantities do not change over time, are only a function of position.

Uniform flow takes place when the cross-section shape of the stream remains constant and is related to uniform motion. Non-uniform flow can be steady (stationary) or unsteady (non-stationary). Non-uniform flow can be divided into slowly varied and rapidly varied (Orzechowski et al. 2001; Hager 2010). The rapidly varied non-uniform flow takes place in situations when

DOI: 10.1201/9781003546436-5

there are considerable differences in depth in channel cross-sections situated close to each other. Such motion is usually caused by local obstructions disturbing the flow and, thus, in relation to a sewage conduit, can be considered as a result of its damage.

Slowly varied non-uniform flow of liquid in channels with a free surface is characterised by the domination of one component of the velocity vector, namely that which is directed along the axis of the channel – the sewage conduit. That property justifies treating that kind of flow as a unidimensional problem (Szymkiewicz 2000; Huisman 2001).

Describing the flow of sewage in a gravitational sewer system, one can make use of the theory of the flow of a liquid with a free surface in pressureless conduits. This is justified by the fact that in closed conduits, but not fully filled with sewage, the flow of sewage takes place solely due to the forces of gravity, inertia, and friction. Movement of this type is caused by the slope, i.e. the longitudinal inclination of the conduit, and is referred to as gravitational flow (Hager 2010).

According to the method of conducting the liquid flow, the following are recognised:

- pressure flow, which occurs in pipes, pressure systems, etc.,
- pressureless flow (with free surface),
- hydraulic jets, which tend to be free or submerged and occur when water is flowing out of pipe or when water is flowing above crest (overflow).

5.2 Steady flow

In order to optimally select the channel diameter (D) for the value of the design flow (Q_{obl}), it is important to correctly determine the gradient of the pipe bottom for the typical ranges of flow velocities (combined sewer $\upsilon = 1–3$ m/s and distribution sewer $\upsilon = 0.8–5.0$ m/s). According to the principles developed in Germany (ATV-A 118 1999), for storm and combined sewers it is recommended to design their cross-section in such a way that the design flux (determined using the method of limiting rainfall intensities) does not exceed 90% of the total capacity; in practice, this corresponds for circular channels to the filling $h/D = 0.75$. This results from the fact that at the filling $h/D \approx 0.75$ the pipe capacity reaches its maximum value (Dąbrowski 2004). In order to ensure continuous ventilation of the pipe and prevent the wastewater from putrefying, it is important to fulfil the above-mentioned assumption in the design of sanitary sewers. When designing pipes, attention should be paid to achieving the minimum flow velocity, which is the condition of "self-cleaning" as well as not exceeding the maximum values (PN-EN 752:2000/2001; Dąbrowski 2004; Hager 2010):

- combined sewer system ($\upsilon_{min} = 1.0$ m/s and $\upsilon_{max} = 3.0$ m/s),
- storm and household ($\upsilon_{min} = 0.8$ m/s and $\upsilon_{max} = 5.0$ m/s).

Kotowski (2011) and Dąbrowski (2004) reviewed the criteria underlying the determination of minimum slope.

Assuming the values of steady-state flow, the volume of wastewater flowing through a circular channel (D), bottom slope (J) with a given filling (h) in the simplest case can be calculated from the Chezy-Manning's formula:

$$Q = F \cdot \sqrt{\frac{8 \cdot g \cdot R_h \cdot J}{\lambda}} = F \cdot \sqrt{\frac{8 \cdot g}{\lambda}} \cdot \sqrt{R_h \cdot J} = F \cdot C(h) \cdot \sqrt{R_h J} \tag{5.1}$$

where: J – slope of the pipe bottom; in practice it is often assumed that the minimum value $J=J_{min}=1/D$; however, it is not a rule, as its value may depend on local conditions (Kotowski 2011), λ – linear resistance coefficient, $C(h)$ – velocity coefficient, R_h – hydraulic radius in the case of circular channel determined from the following formula:

$$R_h = \frac{D \cdot (\varphi - \sin\varphi)}{4 \cdot \varphi} \tag{5.2}$$

F – cross-sectional area (see Figure 5.2.) calculated from the following relation:

$$F = \frac{D^2}{8}(\varphi - sin\varphi)$$

$$\sin\varphi = \frac{4 \cdot |2 \cdot h - D| \cdot \sqrt{h \cdot D - D^2}}{D^2} \tag{5.3}$$

Assuming that conditions of turbulent flow are found in the pipe, which is equivalent to having the value of the velocity coefficient described by Manning's equation, $C = \frac{1}{n} \cdot R_h^{1/6}$ the following relation can be written:

$$Q = \frac{D^{8/3} \cdot (\varphi - sin\varphi)^{5/3}}{2^{13/3} \cdot \varphi^{2/3} \cdot n} \cdot J^{0.5} \tag{5.4}$$

At the stage of modelling or designing stormwater pipes, it is absolutely necessary to remember that for the identification of flow conditions in the pipe (assessment whether turbulent flow is present), the Reynold's number alone is insufficient ($\mathrm{Re} = \frac{4 \cdot \upsilon \cdot R_h}{\nu}$, where: ν – kinematic coefficient of viscosity), but the relation $\lambda=f(\mathrm{Re}, \varepsilon=0.25 \cdot k \cdot R_h^{-1})$ and meeting the condition λ=constant is required. In the case where $\lambda=f(\mathrm{Re}, \varepsilon)$=var, the Coolebroke-White equation is used to calculate the linear drag coefficient:

$$\frac{1}{\sqrt{\lambda}} = -2 \cdot \log\left(\frac{2.51}{\mathrm{Re}\sqrt{\lambda}} + \frac{k}{14.84 \cdot R_h}\right) \tag{5.5}$$

For turbulent flow, the flow conditions in the pipe are shaped by roughness and λ=constant, which leads to the following relationship

$$\frac{1}{\sqrt{\lambda}} = -2 \cdot \log\left(\frac{k}{14.84 \cdot R_h}\right) \tag{5.6}$$

Assuming partial filling of the pipe ($h/D<1$) as well as conditions of steady-state flow and $\lambda=f(\mathrm{Re}, \varepsilon)$=var, the flow rate in the pipe can be calculated from the formula

$$Q=-1.104\cdot D^{5/2}\cdot\frac{(\varphi-\sin\varphi)^{3/2}}{\varphi^{1/2}}\cdot\log\left(\frac{2.51}{\mathrm{Re}\sqrt{\lambda}}+\frac{k\cdot\varphi}{3.71\cdot D\cdot(\varphi-\sin\varphi)}\right)\cdot\sqrt{J} \tag{5.7}$$

Assuming turbulent flow after appropriate transformations (Eq. 5.6), the flow rate in a partially filled pipe can be determined from Eq:

$$Q=-1.104\cdot D^{5/2}\cdot\frac{(\varphi-\sin\varphi)^{3/2}}{\varphi^{1/2}}\cdot\log\left(\frac{k\cdot\varphi}{3.71\cdot D\cdot(\varphi-\sin\varphi)}\right)\cdot\sqrt{J} \tag{5.8}$$

The throughput of a completely filled pipe for the case $\lambda=f(\mathrm{Re},\varepsilon)=\mathrm{var}$ can be determined from the following formula:

$$Q=-6.957\cdot D^{3/2}\cdot\log\left(\frac{0.567}{D\cdot\sqrt{D\cdot J}}+\frac{k}{3.71\cdot D}\right)\cdot\sqrt{J} \tag{5.9}$$

In turn, when $\lambda=\mathrm{constant}$, the pipe throughput is described by the relation:

$$Q=-6.957\cdot D^{3/2}\cdot\log\left(\frac{k}{3.71\cdot D}\right)\cdot\sqrt{J} \tag{5.10}$$

Under operational and design conditions, an important role is played by the improvement in the selection of roughness values, which, in existing stormwater networks, is dependent on the pipe material and local losses generated at the pipe joints, inaccuracies in the axial alignment of the pipes, changes in the slope of the channel bottom, incompletely circular cross-section of the channel, and the effects of aging processes. Therefore, the linear resistance coefficient finally takes the form:

$$\lambda_e=\lambda+\frac{4\cdot R_h}{L}\cdot\sum\xi \tag{5.11}$$

where: λ – linear resistance coefficient, R_h – hydraulic radius calculated from equation (Eq. 5.2), L – pipe length, ξ – local resistance (minor losses) coefficient.

Assuming turbulent flow in the pipe, the flow rate in the pipe calculated with the Chezy-Manning's and Colebrooke-White equations can be compared, obtaining the relation:

$$n_e=\left(4\cdot\sqrt{g}\cdot\left(\frac{8}{R_h}\right)^{1/6}\cdot\log\left(\frac{14.84\cdot R_h}{k_e}\right)\right)^{-1} \tag{5.12}$$

where: k_e – operating (pipe wall) roughness (mm). According to Hager (2010), an average operating roughness value of 0.25 mm is recommended for the design of transit pipes (operating under pressure); this does not take into account local losses. On the other hand, for the dimensioning of pipes with partial filling, it is recommended to assume: k_e = 0.50 mm for sewer collection pipes/channels with sumps up to half the height of the channel cross-section; k_e = 0.75 mm for sewer collection pipes/channels with sumps up to half the height of the channel cross-section; k_e = 1.5 mm for sewer collection pipes/channels with sumps up to half the height of the channel cross-section.

Equation (5.4) and consequent equations given above take into account the fact that along the perimeter of the channel, the roughness coefficient takes a constant value or it is a weighted average of the sections along the perimeter (Kubrak 1998):

$$n = \frac{\sum_{i=1}^{n} n_i \cdot L_i}{\sum_{i=1}^{n} L_i} \tag{5.13}$$

for which particular values of n_i or roughness may be determined (Dąbrowski 2006). However, this approach has a fundamental disadvantage, since the variation in the value of the linear resistance coefficient, which is characteristic for the respective value of n_i, is omitted. This assumption affects the determined variation of $Q=f(h)$, $\lambda=f(h)$ as well as the hydraulic conditions in the pipe. A similar issue occurs when determining the throughput of open channels. In this case, in order to determine the variation of the value of the coefficient of linear resistance along the circumference, an appropriate calculation procedure has to be applied – the method of averaging the coefficient of linear resistance (Kubrak 1998). For this purpose, flow conditions were considered for a fully filled channel with diameter D, bottom slope equal to J, as well as roughness k_1 and k_2 at perimeter lengths L_{01} and L_{02}, respectively (Figure 5.1).

In the linear averaged drag coefficient method, the hydraulic conditions in the cross-section are assumed to be shaped by the relative roughness values and Reynolds number values. Thus, the cross-section (F) through which stormwater flows can be divided into i - parts with area F_i

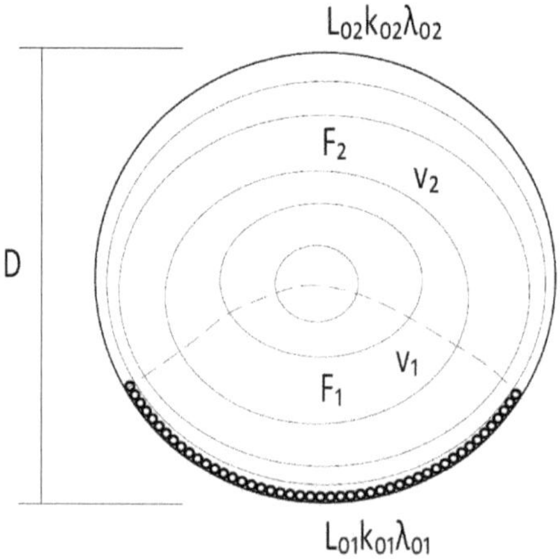

Figure 5.1 Channel throughput calculation scheme using averaged linear drag coefficient method.

where hydraulic conditions are diversified. In these cross-sections, it is assumed that the average velocity values $\upsilon_1=\upsilon_2=\upsilon$ are identical, which means that the hydraulic gradients are also equal to each other $J_1=J_2=J$. The division of the cross-section is carried out in such a way that there are no tangential stresses along the division lines, which means that the division lines should be orthogonal to the velocity isolines in the cross-section.

Assuming Manning's formula in the description of hydraulic phenomena in the channel, it can be written that the average values of flow velocities in particular i-th sections are identical:

$$\upsilon_i=\upsilon_1=\upsilon_2=\upsilon=\frac{1}{n_1}\cdot\left(\frac{F_1}{L_{01}}\right)^{2/3}\cdot J^{0.5}=\frac{1}{n_2}\cdot\left(\frac{F_2}{L_{02}}\right)^{2/3}\cdot J^{0.5}=\frac{1}{n_i}\cdot\left(\frac{F_i}{L_{0i}}\right)^{2/3}\cdot J^{0.5}=\frac{1}{n_{sr}}\cdot\left(\frac{F}{L_0}\right)^{2/3}\cdot J^{0.5} \quad (5.14)$$

Knowing that the sum of the areas F_i, where the flow conditions shaped by the individual roughness (k_i), is equal to the cross-section F, the relation can be written:

$$F=\sum F_i=\sum\frac{n_i^{3/2}}{J^{3/4}}\cdot L_{0i}\cdot\upsilon_i^{3/2}=\frac{\upsilon_i^{3/2}}{J^{3/4}}\cdot\sum n_i^{3/2}\cdot L_{0i} \quad (5.15)$$

Expressing the average flow velocity in the cross-section as a function of the average value of the roughness coefficient (n_{sr}), it can be written that the following relation occurs

$$n_{sr}=\left(\frac{\sum n_i^{3/2}\, L_{0i}}{L_0}\right)^{2/3} \quad (5.16)$$

In relation (5.13), expressing the flow velocity with the general equation $\upsilon=f(R_h, \lambda)$ yields:

$$\upsilon_i=\upsilon=\upsilon_1=\upsilon_2=\sqrt{\frac{8\cdot g\cdot R_{h1}\cdot J}{\lambda_1}}=\sqrt{\frac{8\cdot g\cdot R_{h2}\cdot J}{\lambda_2}}=\sqrt{\frac{8\cdot g\cdot R_h\cdot J}{\lambda_{sr}}} \quad (5.17)$$

and, after appropriate transformations

$$\frac{R_{h1}}{\lambda_1}=\frac{R_{h2}}{\lambda_2}=\frac{R_h}{\lambda_{sr}} \quad (5.18)$$

where: R_h – hydraulic radius of the cross-section through which stormwater flows.

Knowing that the cross-section of the stream through which the stormwater flows (F) is the sum of the partial surfaces (F_i), where the flow conditions are shaped by roughness k_i, the following relationship can be written based on Eq. 5.17:

$$F=\sum F_i=\sum\frac{\upsilon_i^2\cdot\lambda_i\cdot L_{0i}}{8\cdot g\cdot J}=\frac{\upsilon_i^2}{8\cdot g\cdot J}\sum\lambda_i\, L_{0i} \quad (5.19)$$

Assuming that the value of flow velocity in the channel cross-section can be expressed in terms of the average value of the linear drag coefficient (λ_{sr}), the equation below is obtained:

$$\upsilon = \upsilon_i = \sqrt{\frac{8 \cdot g \cdot R_h \cdot J}{\lambda_{sr}}} = \sqrt{\frac{8 \cdot g \cdot J \cdot \sum F_i}{\lambda_{sr} \cdot \sum L_{0i}}} = \sqrt{\frac{8 \cdot g \cdot J \cdot \frac{\upsilon_i^2}{8 \cdot g \cdot J} \sum \lambda_i \cdot L_{0i}}{\lambda_{sr} \cdot \sum L_{0i}}} = \upsilon_i \cdot \sqrt{\frac{\sum \lambda_i \cdot L_{0i}}{\lambda_{sr} \cdot \sum L_{0i}}} \tag{5.20}$$

Knowing that $\upsilon = \upsilon_1 = \upsilon_2 = \upsilon_i$, on the basis of Eq. 5.19, the following formula is obtained:

$$\lambda_{sr} = \frac{\sum \lambda_i \cdot L_{0i}}{\sum L_{0i}} \tag{5.21}$$

In the method described above, the values of hydraulic radii R_{hi} are determined iteratively, and the criterion for their termination is the equality of the values of linear resistance coefficients determined in successive iterations. The advantage of the approach discussed above is that the value of linear resistance coefficients can be determined at the circumference of the channel, which is not taken into account in the classical approach.

5.3 Unsteady flow

5.3.1 *Unsteady open channel flow*

One of the important determinants of the variability of the value of flows in the cross-section closing the analysed catchment is the system of the stormwater level in the sewer network. Its temporal variability is determined by the level of stormwater inflow into the sewer conduits and its distribution along their length, as well as conduit diameters, roughness of sewer walls, and conduit slopes (Zhou et al. 2019; Szeląg et al. 2021b). Rainfall phenomena constituting the basis for dimensioning the stormwater sewer system (combined) are intense rainfall events, determining the high variability of the flow value per time unit (Guo et al. 2021; Addison-Atkinson et al. 2022). Therefore, the phenomena accompanying the sewer network flow cannot be modelled as steady flow conditions. In these cases, the flow phenomena that may occur in the sewer network have an unsteady character. The theoretical basis for the description of unsteady flow in the conduit and lateral sections is illustrated by the Saint Venant differential equation system, based on the following equation of continuity of flux and momentum:

$$\frac{\partial U}{\partial t} + \beta U \frac{\partial U}{\partial x} + g \frac{\partial}{\partial x}\left(H + z_d\right) + \frac{gn^2}{H^{4/3}}\left|U\right|U + \frac{1}{\rho}\frac{\partial p_a}{\partial x} - \frac{a}{\rho H}\left|W\right|W - \frac{\eta^D}{\rho}\frac{\partial^2 U}{\partial x^2} = 0 \tag{5.22}$$

$$\frac{\partial H}{\partial t} + \frac{\partial}{\partial x}\left(UH\right) = 0 \tag{5.23}$$

The first two terms of the first equation describe the force of inertia, the third – the force of gravity and pressure, the fourth – the force of friction against the bottom, the fifth – atmospheric pressure, the sixth – the force of friction caused by air flow, and the seventh – dispersion of momentum.

In many cases encountered in practice, the final three terms of equation (Eq. 5.22) are omitted. This is possible, as its analysis in a dimensionless system reveals a small numerical value of the term of longitudinal dispersion of momentum, compared to the other terms. In most cases, also, changes in air pressure and in its flow in sewer systems are negligibly small; they may prove to be significant only in rare situations of nearly complete filling of a sewage conduit or in the case of occurrence of the so-called chimney draught. Moreover, assuming coefficient $\beta = 1$ and introducing in the formula the slope of the channel s, as well as a drop in the energy line S (friction slope), a notation is obtained which, together with equation (Eq. 5.23), forms the Saint Venant systems of equations:

$$\frac{\partial U}{\partial t} + U\frac{\partial U}{\partial x} + g\frac{\partial H}{\partial x} = g\left(s - S\right) \tag{5.24}$$

$$\frac{\partial H}{\partial t} + \frac{\partial}{\partial x}\left(UH\right) = 0 \tag{5.25}$$

The S parameter included in the above set of equations, describing waves propagation in open channels at their unit width, may be written as follows:

$$S = \frac{n^2}{H^{4/3}}\left|U\right|U \tag{5.26}$$

However, in engineering practice, other independent variables are commonly used. The flow rate and free surface coordinates are used instead of flow velocity and sewage depth. Considering the influence of velocity distribution diversity at cross-section β, and marking the cross-section area as A, flow rate as Q, one must notice that the considered variables refer to the whole cross-section of a pipe and are properly interpreted. Thus, U is a mean flow velocity at a given cross-section and S is drop in energy line reflected to the whole cross-section – connected to the hydraulic radius R_h and roughness coefficient n of the material of the inner side of the pipe walls, which may be presented as (Szymkiewicz 2000; Zoppou 2001):

$$U\left(x,t\right) = \frac{1}{A}\iint_A u\left(x,y,z,t\right)\,dA = \frac{Q}{A} \tag{5.27}$$

$$S = \frac{n^2}{R_h^{4/3}}\left|U\right|U \tag{5.28}$$

where: R_h – hydraulic radius, which is a relation of cross-section area A to the wetted perimeter (m), O_z – the peripheral length of contact between transported sewage and pipe material (m) (Hager 2010).

Coefficient β is sometimes defined as the ratio of the square of stream momentum at the actual distribution of velocity to the square of momentum at averaged velocity in the cross-section considered:

$$\beta = \frac{\iint_A u^2 \, dA}{U^2 A} \tag{5.29}$$

Assuming additionally that the level of the channel bottom does not change in time, an alternative form of the system is obtained (Eqs. 5.24 and 5.25), which is frequently employed for the description of unsteady flow in open channels (Hager 2010):

$$\frac{\partial Q}{\partial t} + \frac{\partial}{\partial x}\left(\frac{\beta Q^2}{A}\right) + gA\frac{\partial h}{\partial x} + g\frac{n^2}{R_h^{4/3}}\frac{|Q|Q}{A} = 0 \tag{5.30}$$

$$\frac{\partial A}{\partial t} + \frac{\partial Q}{\partial x} = 0 \tag{5.31}$$

where: R_h – hydraulic radius (m), A – stream cross-sectional area (m^2), Q – flow rate (m^3/s), β – dimensionless velocity coefficient (-), n – roughness coefficient (s·m$^{-1/3}$).

The systems of equations presented above, (Eqs. 5.24 and 5.25) and (Eqs. 5.30 and 5.31), are referred to as the complete dynamic model or the dynamic wave model (Szymkiewicz 2000; Zoppou 2001). As demonstrated by numerous experiments described in the literature on the subject, it permits obtaining results with good agreement with measured data in cases when the assumptions adopted during the derivation of the equations are met, and namely:

- the motion of the fluid is a slowly varied flow,
- the force of gravity is the only mass force,
- the slope of the channel bottom is small,
- the vertical distribution of velocity is treated as uniform,
- there is no side inflow or tributary on the section of the sewage conduit under consideration,
- the fluid described in incompressible and homogeneous – with no variation of density.

The equations denoted as (Eqs. 5.24 and 5.25) and (Eqs. 5.30 and 5.31) are systems of first-order partial differential equations, of the hyperbolic type, and describe slowly varied unsteady motion. For the solution of a system of equations, of this kind it is necessary to define in advance the initial and boundary conditions, which is usually achieved by assuming that at the initial moment the flow was steady – this permits the calculation of the initial values of $z(x)$ and $Q(x)$ for time $t=t_0$ (Hager 2010). In the case on hand, the boundary conditions relate most often to the values of $z(t)$ and $Q(t)$ at the initial and final cross-section of the sewage conduit, respectively.

The systems of equations discussed, in conjunction with the boundary conditions, are usually solved numerically – e.g. by means of the finite differences or finite elements methods. The full Saint Venant equation should be used in cases when the Froude number for the system under analysis is smaller than unity (Fr < 1). More detailed information on the criterion Froude number is given in the next subchapter. The basic form of the Saint Venant equation (5.28, 5.29) and some of its reduced versions require the determination of the value of drop of the energy line S which can be calculated as in formulae (Eqs. 5.26 and 5.28) and in the formulae presented below.

The empirical equation for the friction slope in unsteady flow can be written in the following manner proposed by Hager (2010):

$$S = \frac{1}{C_{\mathrm{CH}}^{2}} \frac{1}{R_h^{2a_e}} U\left|U\right| \tag{5.32}$$

where: S – friction slope (m/m), C_{CH} – Chezy empirical roughness coefficient ($m^{1/2} \cdot s^{-1}$), a_e – empirical exponent (-).

The empirical roughness coefficient may be presented as the Darcy-Weisbach equation:

$$C_{\mathrm{CH}} = \left(\frac{8g}{\lambda}\right)^{1/2} \tag{5.33}$$

Therefore, in the flow model it is possible to employ the equation that is encountered in the literature under the name of the Darcy-Weisbach equation:

$$S = \frac{\lambda}{8g} \frac{1}{R_h} U\left|U\right| \tag{5.34}$$

where: U – mean axial velocity in the active cross-section (m/s), λ – dimensionless coefficient of friction factor (-).

Coefficient λ is related to internal friction and to the contact surface of flow turbulence with the conduit walls, and more accurately with their roughness. The kind of material and the state of the surface of the conduit are characterised by the sand roughness index studied in greater depth by Kubrak (1998), Huisman (2001), Akan & Houghtalen (2003) and Hager (2010), who proposed the following classification of surface:

$$(\mathrm{smooth}) \quad 4 < k^{*} = \frac{k\,U_{*}}{\nu} < 100 \quad (\mathrm{rough}) \tag{5.35}$$

where: k^* – dimensionless coefficient of surface roughness (-), k – equivalent sand roughness, defined as the size of sand grains that cause the same flow resistance as that observed in experimental measurements in a given conduit (m), ν – kinematic viscosity, adopted for sewage to have the value of $1.31 \cdot 10^{-6}$ (m^2/s), U_* – dynamic velocity (m/s).

The coefficient of linear hydraulic resistance λ is calculated for the following cases (intervals) (ATV-DVWK-A110P 1988; Hager 2010):

- turbulent flow in hydraulic-smooth conduits – acc. to Prandtl,
- completely turbulent flow at perfect roughness – acc. to Prandtl,
- flow in conduits with natural technical roughness – acc. to Colebrook.

For calculations of sewage channels and conduits the last of the cases listed has an application, and the equation that describes it is encountered in the literature under the name of the Prandtl-Colebrook formula (ATV-DVWK-A110P 1988; Hager 2010), or the Colebrook-White equation.

The dimensionless friction factor can be calculated after the Prandtl-Colebrook formula:

$$\frac{1}{\sqrt{\lambda}} = -2\log\left(\frac{2.51}{\mathrm{Re}\sqrt{\lambda}} + \frac{k/D}{3.71}\right) = -2\log\left(\frac{2.51}{\mathrm{Re}\sqrt{\lambda}} + \frac{1}{3.71} + \frac{k}{4R_h}\right) \tag{5.36}$$

where: D – pipe diameter (m), k – wall sand roughness (m), Re – dimensionless Reynolds number (-).

$$\mathrm{Re} = \frac{\rho U D_z}{\eta} = \frac{U\, 4R_h}{\nu} \tag{5.37}$$

where: ρ – density (kg·m³), η – coefficient of dynamic viscosity (Pa·s), D_z – equivalent diameter of the conduit (m), D_z=4 R_h for most non-circular conduits (m).

The dimensionless coefficient of linear resistance can be described with the formula:

$$\lambda = \left[-2\log\left(\frac{14.84\, R_h}{k}\right)\right]^{-2} \tag{5.38}$$

Another method of calculation of the slope of the energy line (3.30) is the application of the Manning's equation to calculate the Chezy coefficient:

$$C_{\mathrm{CH}} = \frac{1}{n_M} R_h^{1/6} \tag{5.39}$$

where: n_M – roughness coefficient acc. to Manning (s·m$^{-1/3}$).

The approach proposed by Manning was derived for steady flow, but it is also used for unsteady flow after taking into account the relations of the energy line and n to time and to the wetted perimeter. The Chezy equation, in which coefficient C_{CH}, the empirical roughness coefficient is represented by the Manning's equation:

$$U = M\, R_h^{2/3}\, s^{1/2} \tag{5.40}$$

$$Q = U\, A \tag{5.41}$$

where: U – mean flow velocity in steady conditions (m/s), R_h – hydraulic radius (m), s – slope of the channel bottom (m/m), A – conduit cross-sectional flow area (m²), M – Manning-Strickler coefficient dependent only on conduit wall roughness (m$^{1/3}$·s^{-1}), values of that coefficient most frequently encountered in sewer systems are given in the Table 5.1.

The Pawłowski equation is another alternative in hydrodynamic calculations, it relates the energy line slope to the roughness coefficient and to the hydraulic radius (Hager 2010):

Table 5.1 Coefficient *n* by Manning, *M* coefficient by Manning-Strickler (DHI 2000; Dąbrowski 2004)

No	*Type of surface, material*	$n = M^{-1}$ ($s \cdot m^{-1/3}$)	M ($m^{1/3} \cdot s^{-1}$)
1	Smooth glaze surfaces	0.009	111
2	Planed wood	0.010	100
3	Smooth concrete	0.0118	85
4	Normal concrete	0.0133	75
5	Rough concrete	0.0147	68
6	Plastic	0.0125	80
7	Smooth stone	0.0125	80
8	Ceramics	0.0143	70
9	Iron	0.0143	70
10	Bricks	0.0167	60
11	Broken stone wall, channel in bad condition	0.020	50
12	Channel in extremely bad condition, silted	0.030	33

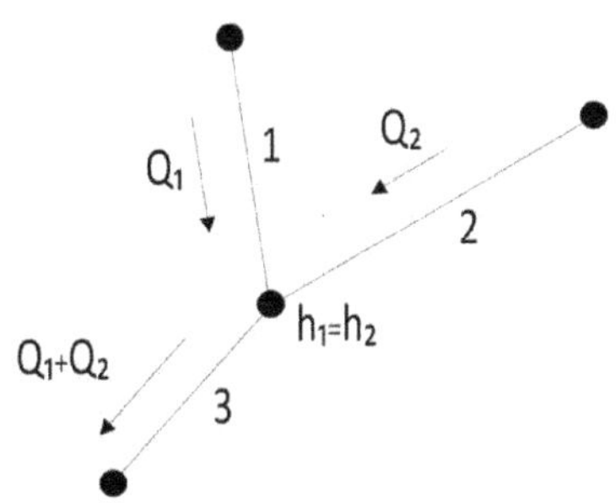

Figure 5.2 Scheme of calculations for channel networks.

$$S = n_M^{\,2} \frac{1}{R_h^{\,2y}} U \left| U \right|$$

$$y = 2.5\sqrt{n_M} - 0.13 + 0.75\sqrt{R_h\left(\sqrt{n_M} - 0.10\right)} \tag{5.42}$$

The equations presented above are the methods most frequently applied in practice for the description of the energy line slope.

It is important for calculations for the sewer network (conduit system) that the filling rate should be identical at the junction points and that the flow in the cumulative collector should be the sum of flows in individual channels (Figure 5.2). An overflow of stormwater to the surface is an exception.

The relationship described by the system of equations (Eqs. 5.30 and 5.31), where all terms are included, is the so-called dynamic wave equation. It facilitates modelling the operation of sewer systems within which there may be changes in the flow direction, pressure flows, stormwater accumulation, and flooding caused by limited gravitational flow or regulation of the flow stream by means of reservoirs, discharges, regulators, etc., are solved at the stage of simulation of operation by application of complex numerical algorithms (Hager 2010), which have a significant impact on the calculation results. Hence, to address the problems with the achievement of reliable simulation results, the system of equations (Eqs. 5.30 and 5.31) could be simplified.

5.3.1.1 *Saint Venant equation approximations*

The simplifications have a substantial impact on the application of the results obtained in this way to assess the operation of the sewer network. At the stage of calculations, the simplification is equivalent to the exclusion of important phenomena (changes in the flow direction, pressure flow, stormwater flooding, local pressure losses, stormwater accumulation) occurring in the sewer system, which largely limits its application.

The dynamic wave equation (Eq. 5.30) is often considered a "complete" equation because it contains all terms describing the dynamic effects of steady flow with a free surface. The solution of this equation, as well as its simplified form together with the continuity equation and the specified initial and boundary conditions, therefore, various simplifications and approximations of these equations are often used (Yen 1986).

Different levels of approximations of the dynamic equation are obtained by neglecting the individual terms of Eq. 5.30. If the acceleration term $\partial Q/\partial t$ in this equation is neglected, the so-called the equation of a quasi-stationary dynamic wave is obtained. The diffusion wave equation is obtained by neglecting another term expressing convective acceleration. In turn, neglecting the term expressing the change in pressure ($\partial h/\partial x$), yields the equation of the kinematic wave.

To explain the individual approximations, let write the Saint Venant equation in the form

Full dynamic wave $$i_e = i_0 - \frac{\partial y}{\partial x} - \frac{v}{g} \cdot \frac{\partial v}{\partial x} - \frac{1}{g} \cdot \frac{\partial v}{\partial t} \tag{5.43}$$

Diffusive wave $$i_e = i_0 - \frac{\partial y}{\partial x} - \frac{v}{g} \cdot \frac{\partial v}{\partial x} \tag{5.44}$$

Kinematic wave $$i_e = i_0 \tag{5.45}$$

5.3.1.2 *Diffusive wave*

The name of this approximation is taken from thermodynamics, where this approximation of the dynamic equation together with the continuity equation is known as the convection diffusion equation. Neglecting the local and convective acceleration members, the diffusion wave approximation requires that the flow change gradually in time and space. Maintaining the pressure term $\partial Q/\partial x$ preserves the possibility of including the backwater effect in the simulation.

Although the accuracy of this approximation is, of course, not the same as that of the complete dynamic equation, the diffusion wave approximation is much simpler and faster, and the above-mentioned backwater effect is very often needed to simulate properly the flow in the sewer systems.

The diffusion wave requires, in addition to the initial, two boundary conditions to meet the condition of unambiguousness of the solution. If backflow occurs during the flow in the drain, this approximation is able to reflect this phenomenon based on the negative values of Q flows.

The diffusion wave equation, derived from the Eq. 5.30, can also be written in the following form:

$$g \cdot F \cdot \frac{\partial h}{\partial l} + g \cdot F \cdot S_f - g \cdot F \cdot J = 0 \tag{5.46}$$

5.3.1.3 *Kinematic wave*

Calculations taking into account the significant variability of the flow parameters in time and along the length of the open channel showed that the first two terms in the equation of momentum conservation, i.e. $\frac{\partial Q}{\partial l}$ and $\frac{\partial}{\partial l}\left(\frac{Q}{F^2}\right)$ have values of 10^{-4} and $1.5 \cdot 10^{-4}$. It can be shown that for the value of $J \approx 10^{-3}$ with slight changes in the filling along the channel length, the values of $\frac{\partial h}{\partial l}$ are relatively small in relation to $g \cdot J$, which indicates that approximated $J = S_f$ can be assumed. These assumptions yield the following system of equations:

$$\frac{\partial F}{\partial t} + \frac{\partial Q}{\partial l} = 0 \tag{5.47}$$

$$S_f = J \tag{5.48}$$

This approximation is called kinematic because it ignores all dynamic effects of flow, except slope. In other words, for each Δt, within the range of the calculated length Δx, the water level is parallel to the bottom of the pipe; thus, the flow is steady. Thus, the kinematic wave does not flatten; it only "slides" along the bottom of the sewer.

The kinematic wave equation requires only one boundary condition in conjunction with one initial condition in order to obtain a simulation solution. This initial condition is usually specified at the upstream end of the section, usually as an inflow hydrogram.

The solution does not respect the downstream boundary condition. This is also the greatest disadvantage of this approximation – the inability to deal with the backwater effect, which is usually a relatively important aspect of flow simulation in sewer networks. In practice, this means that the kinematic wave approximation can only be applied in cases where the flow is independent of the conditions at the downstream end of the sewer section, i.e. if $Fr \geq 1$. Neglecting the effects of pressure and acceleration, the kinematic wave does not have a "mechanism" to flatten the flow peaks, which occurs in real situations.

5.3.1.4 *Steady flow conditions*

Another simplification of the Saint Venant equation system (Eq. 5.47, Eq. 5.48) is the assumption of steady flow conditions, where the volume of the stormwater flow is constant along the channel length and the stormwater table along the channel length is parallel to the bottom. This model is described by

$$\frac{\partial Q}{\partial l} = 0 \tag{5.49}$$

$$S_f = J \tag{5.50}$$

The aforementioned system has a wide application in engineering practice as a basis for designing channel diameters.

5.3.1.5 *Comparison of the results using Saint Venant equation approximations*

A comparison of the results obtained using the individual approximations of the Saint Venant equations is shown in Figures 5.3 and 5.4. All examples were recalculated using the MOUSE simulation model (DHI 2000).

Figure 5.3 shows the resulting hydrograms after flow transformation in a pipe with a diameter of 1.40 m, a slope of 1‰, and a length of 1,000 m, with a hydraulic roughness factor of 0.013 (according to Manning). As can be seen from the figure, there are no significant differences between the diffuse and dynamic waves, but the kinematic wave gives significantly different results.

Figure 5.4 shows the calculated level curves at the occurrence of backwater effect (flow obstruction at the end of the sewer) and at the hydraulic drop (free discharge at the end of the

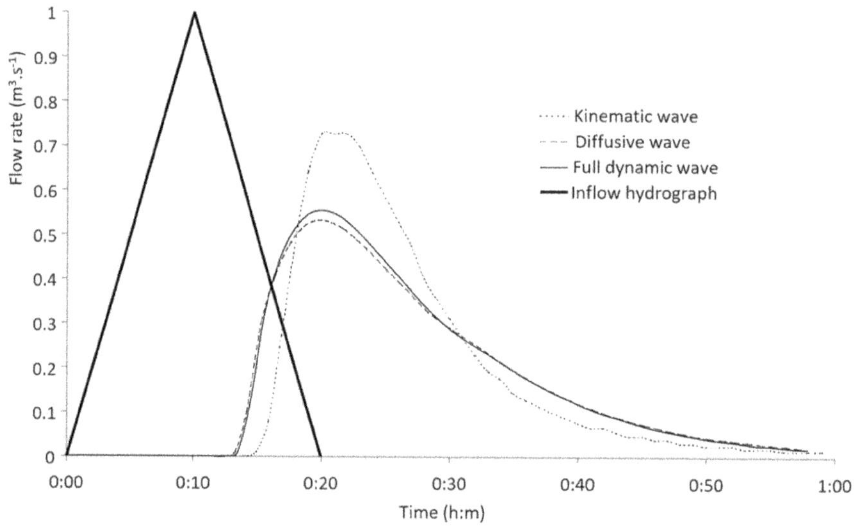

Figure 5.3 Comparison of results using different Saint Venant equation approximations.

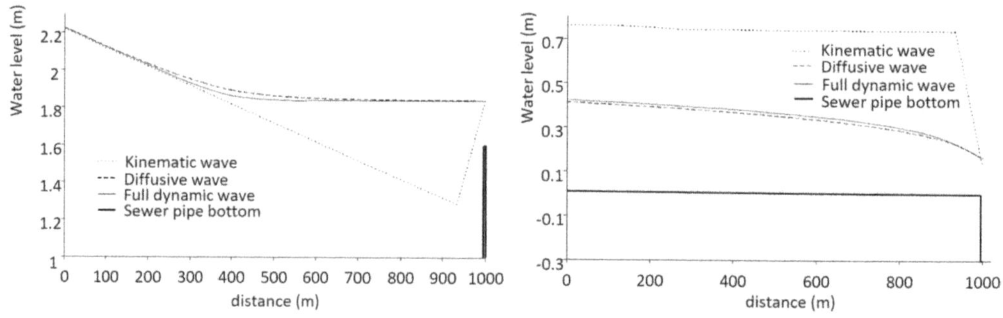

Figure 5.4 The course of water levels during backwater and free outflow at the end of the sewer.

sewer). As it can be seen from the figures, the kinematic wave approximation almost completely fails in these cases, while the diffusion and dynamic waves give almost identical results.

5.3.1.6 *Unsteady pipe (pressurised) flow*

When exceeding the pipe capacity of the free-surface flow (i.e. at pressure flow), the flow area is constant, equal to the entire cross-section profile, i.e. the term $\partial S/\partial x$ is equal to zero. Then the system of Saint Venant's equations (Eqs. 5.30 and 5.31, continuity and momentum equation) can be written in the following form:

$$Q = S_k v \tag{5.51}$$

$$\frac{1}{g}\frac{\partial v}{\partial t} + \frac{\partial}{\partial x}\left[\frac{\beta v^2}{g} + \frac{P_a}{\gamma}\right] = i_e + \frac{1}{\gamma S}\frac{\partial F_t}{\partial x} \tag{5.52}$$

where: P_a – piezometric flow pressure.

If the flow profile has a constant cross-section and an incompressible liquid flows in it in the entire flow cross-section along the entire length, then $\partial v/\partial x = 0$. If the spatial changes β a F_t are neglected by integrating the Saint Venant's momentum equation (Eq. 5.30) along the length of the pipe L obtain:

$$\frac{L}{gS}\frac{\partial Q}{\partial t} = H_v - H_0 - \left[\zeta_v + \zeta_o\right]\frac{Q^2}{2gS_k^2} i_e L \tag{5.53}$$

where: H – level at the inlet to the pipe, H_0 – level at the outlet of the pipe, ς_v and ς_o – inlet and outlet local loss coefficient according to Figure 5.5.

When simulating the flow in a pipe, the equations are initially used to calculate the flow with a free surface, because the pipe is usually not surcharged. However, when the pipe is full during the simulation, and identified according to a certain criterion as a pipe pressure flow, the simulation starts to use the equations for calculating the pressure flow, e.g. Eqs. 5.51 and 5.53. Also, if there is a decrease in flow and a pressureless flow occurs in the sewer according to the criteria,

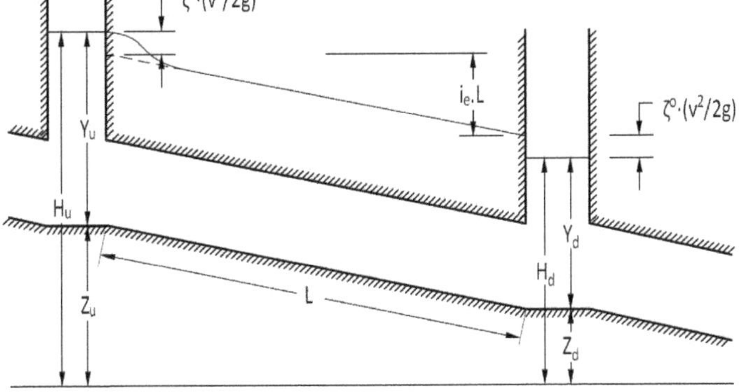

Figure 5.5 Hydraulic concept of pressure flow calculation.

the equations (Saint Venant's equations) for surface free flow will be used again. This change in the use of equations of one or the other type does not cause major problems in programming, but it is a relatively difficult matter, especially due to numerical instabilities and the occurrence of hydraulic unstable phenomena in the transition from non-pressurised to pressurised flow and vice versa.

Because the knowledge of flow type change is somewhat limited, there are many ambiguities and conjectures in determining the flow type transition criterion. The usually specified criteria are the depth, or the submerged pipe inflow or outflow. The best-known criterion is the hydraulic surcharge of the pipe, which is, however, calculated as a steady, uniform flow. If this flow rate is exceeded, it will be used in the calculation of the pressure flow equation, and vice versa, if the calculated flow rate is less than the capacitive flow rate, the pressureless flow equations are used.

To reduce the number of problems in the calculation of pressure flow in the sewers, Preissmann developed the theory of the so-called open slot (Hager 2010). The idea of this theory is to transform the pressure flow into a theoretical free surface flow. This is created by introducing a continuous, hypothetical, narrow, piezometric slot located at the top of the drain along the entire length of the drain, as shown in Figure 5.6.

It is assumed that the slot is narrow so as not to cause an error in the flow volume; however, the slot, on the other hand, must not be very narrow in order to avoid numerical problems in connection with the rapidly changing pressure head at inflows. The theoretical basis for determining the slot width is that the velocity of the pressure wave in the pipe is taken into account,

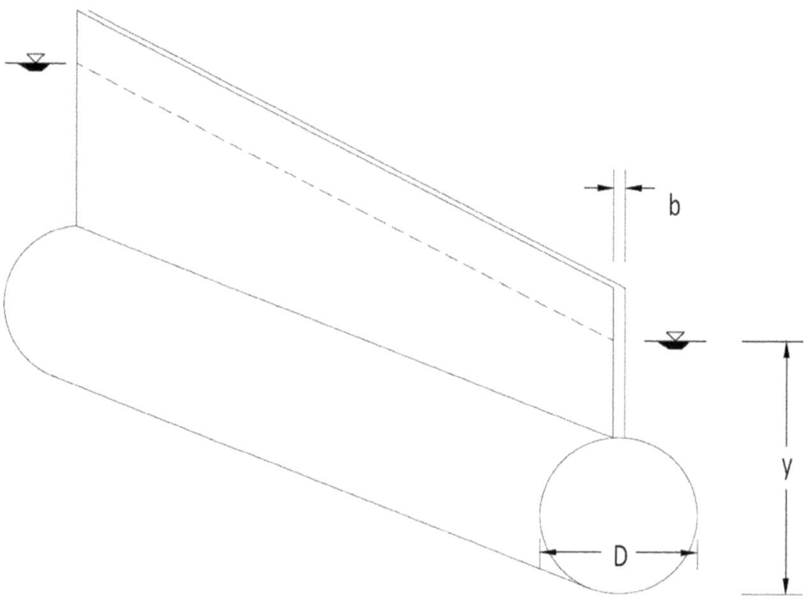

Figure 5.6 Concept of the Preissmann slot.

considering the compressibility of the fluid and the elasticity of the pipe. The width of the slot should be such that the velocity in the slot is the same as in the pipe. The speed of the wave c_1 in the pipe with the slot is:

$$c_1 = \sqrt{\frac{gS}{b}} \tag{5.54}$$

where: b – width of the slot, S – flow area.

Neglecting the increase of the area by the slot, i.e. by assuming that $S=S_k=\pi D^2/4$ for circular cross-sections, and finally by assuming that the velocity c_1 is the same as the velocity of the pressure wave in an elastic pipe without a slot, the equation for the theoretical slot width is obtained.

$$b = \frac{\pi g D^2}{4c_a^2} \tag{5.55}$$

The use of a hypothetical Preissmann slot in the simulation of pressure flow in sewer pipes has the following advantages:

a it uses only Saint Venant equations and there is no need to change the type of equations when changing the flow type; no other approach is needed under changed boundary conditions,
b it is not necessary to define the flow type selection criterion,
c it is not necessary to have an overview of the type of flow in the sewers in different time steps during the calculation,
d it allows the section-by-section calculation procedure as in pressureless flow and can be used for a situation where the pressure flow is only in the sewer part,
e it requires fewer assumptions than the standard approach to maintain numerical stability,
f it is easier to program.

However, it is necessary to mention some disadvantages of this method:

g it causes problems with calculation accuracy at volumes and momentums if the slot is too wide, stability problems if the slot is too narrow,
h it is necessary to use two equations in the calculation (Saint Venant's momentum and continuity equations), while in the classical approach one equation is sufficient for the simulation,
i the whole solution is hypothetical, rather than real (Yen 1986; Bartušková 1985).

5.4 Numerical solutions of Saint Venant equations

In this chapter, some methods for solving partial differential equations will be described. There are several methods for solving this kind of equation, but the methods used to solve Saint Venant's equations in practice will be the main focus. One of the methods is to determine the exact solution in analytical form. However, apart from the simplest cases, this solution cannot be determined, so approximate methods have been developed to solve the problems for linear partial differential equations. The most commonly used are numerical methods that give a solution at discrete points of the area, i.e., in tabular form. These methods are particularly advantageous when using a computer. The most widespread numerical methods for solving partial differential equations in practice are the so-called mesh methods. The most popular mesh methods are the finite difference methods (Miller & Cunge 1975; Singh 1996), and there are many methods and schemes for solving the system of partial differential equations. The most popular are the explicit and implicit methods, and the Crank-Nicolson method. Besides the finite difference methods, there are also the finite element methods, widely used in engineering and computational fluid dynamics (CFD) (Chow 1959; Hirt & Nichols 1981; Chen et al. 2002; Szymkiewicz 2010; Ghanem 2020).

The theoretical procedure of the mesh method is as follows (Šalát 1981):

In a bounded region G with boundary ∂G, in which a solution of a given boundary value problem for partial differential equations is sought, a set of points S (where $X_k \in G \cup \partial G$) isolated from each other is chosen (Figure 5.7). The set S is called a network, and its elements – points X_k – nodes of the network. If node $X_i \in \partial G$, it is called a boundary node. The other nodes of the network are internal nodes.

The solution of the given problem will be sought only in the nodes of the network S, by replacing the derivatives occurring in the partial differential equations and in the marginal, respectively, under the initial conditions by suitable linear combinations of approximate solution

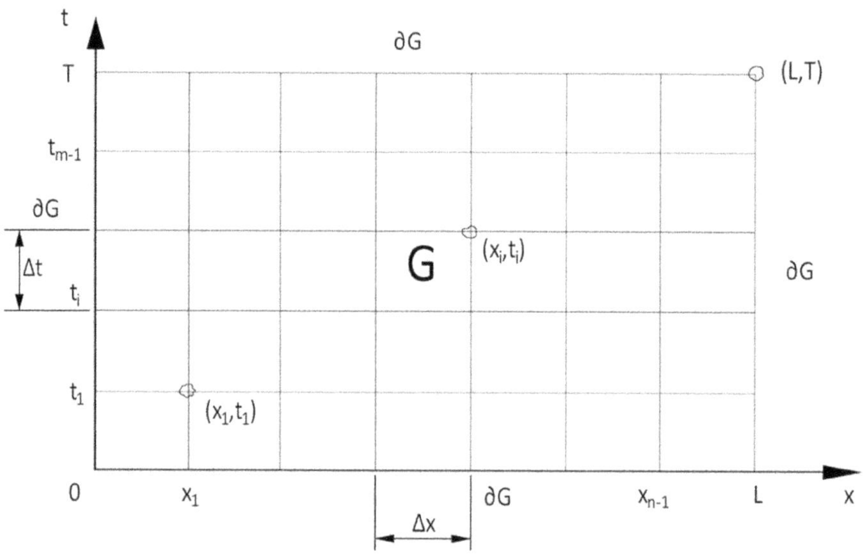

Figure 5.7 Concept of the mesh for the finite difference methods.

values in the network nodes. If it is possible to calculate the value of an approximate solution in a node on the basis of known values according to some explicit formula, this is an explicit method.

If this option is not available and all the unknowns in one section of the network have to be calculated at once, then this is the default method. However, it should be borne in mind that as the number of section divisions increases, so does the number of equations that need to solve in such a procedure. It is assumed that the number of arithmetic operations in solving a system of equations increases with the third power of the number of equations (with the "classical" Gaussian elimination method, although there are already more efficient algorithms, it is not clear that they are faster). When the network is "compacted", there is a large increase in the number of arithmetic operations; e.g., with a double division of the network, the number of operations (and thus the time required for the calculation) increases not twice, but eight times (i.e., there is a cubic dependency).

The solution of unsteady fluid flow in the pipe is a task leading to the solution of a system of partial differential equations. Here, Area G represents a rectangle with dimensions $L \times T$, where L is the length of the pipe and T is the time at which the calculation is to end (otherwise, this time is not limited by anything).

Thus $G = \{(x,t)\} : 0 \le x \le L,\ 0 \le t \le T$, under the initial condition $u\ (x, t) = f(x)$ in time $t = 0$ and with boundary conditions $u\ (x, t) = g(t)$; $x = 0$, $u\ (x, t) = q(t)$; $x = L$ and with the initial condition $u\ (x, t) = w(t)$; $t = 0$.

The network S can be selected in different ways in individual cases (depending on the shape of the area G or the type of task), but it should best approximate the area G. As can be seen in Figure 5.3, an uneven rectangular network is usually chosen. By choosing a rectangular grid, the difficulties that arise when the boundary (or part of the boundary) ∂G is curvilinear are avoided. Then, the derivatives occurring in the boundary conditions have to be approximated by means of special differential relations, which are usually less accurate than the differential relations used in the internal nodes of the network. Similar difficulties are caused by the discontinuity of marginal, respectively, initial conditions.

The essence of finite difference methods is the replacement of differential terms in equations by difference terms in a selected network of points in the space-time domain. This process is called discretisation. The values of the variables are then determined numerically at the computational points of this space-time network in discrete steps Δt and Δx. The finite difference method is based on Taylor's expansion (Taylor's series) of the function f:

$$f(x) = \sum_{n=0}^{\infty} \frac{f_{(a)}^{(n)}}{n!} (x-a)^n \tag{5.56}$$

where: a – point at which the value of the function $f(a)$ is known, x – point at which the value of the function f is to be calculated.

The symbol $f_{(a)}^{(n)}$ is understood as the nth derivative of the function f at the point a. However, this assumes a continuity of derivatives of the function up to the nth order. For example, for the simplest expansion of a function (two expansion numbers are considered, i.e. $n = 0$ and $n = 1$) can be obtain:

$$f(x) = f(a) + f'_{(a)} (x-a) \tag{5.57}$$

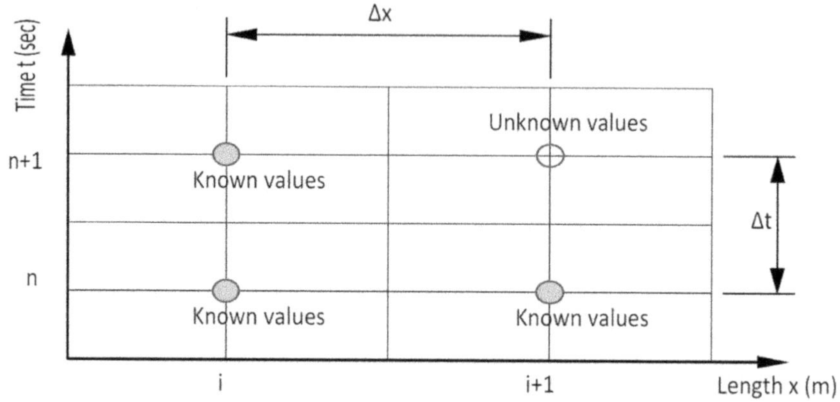

Figure 5.8 Concept of the explicit scheme.

5.4.1 *Explicit schemes*

In explicit schemes (Figure 5.8), Saint Venant equations are replaced by a set of equations in the form of finite differences, so that it is possible to express the value of unknown variables in one timestep from the values of known variables, or from the values of already calculated variables. Depending on the number and position of the points used to express the finite differences by which the differential terms are approximated, the explicit schemes can be divided into many groups. The most popular are, e.g., the Lax-Wendroff scheme, the Lax scheme and various four- and six-point schemes.

Explicit schemes are conditionally stable and in their application it is necessary to keep the Courant's stability condition (Eq. 5.60), which is a necessary and sufficient condition for the stability of explicit calculation schemes.

Explicit schemes are relatively easy to understand, to formulate, and to create programs. However, they are relatively inefficient in calculation, because they have problems of numerical instability of the calculation. Also, the result of the whole calculation is very dependent on the values of Δx and Δt, which is also related to the question of the accuracy of the calculation. The exact value of the solution is usually unknown, so it is questionable which solution is actually correct (e.g., when using different Δx and Δt). To minimise numerical instability, it is necessary to select the calculation steps Δx and Δt so that they meet the Courant stability criterion. However, the required Δt is often so small that even on fast and powerful computers, the calculation takes a relatively long time, while the accuracy of the calculation is still not ensured. From a practical point of view, the use of explicit calculation schemes in models carries many limitations and questions.

5.4.2 *Implicit schemes*

Implicit schemes express unknown values of variables implicitly by a set of algebraic equations, in which differential terms are replaced by difference terms, and by solving this set of equations for a whole time step at a time (Marsalek et al. 1998). By substituting the differential relations for the variables Q and h into the Saint Venant equations, we obtain two algebraic equations for each node of the time-space network.

$$\mathbf{C}_i \left(Q_{i+1,n+1}, Q_{i,n+1}, h_{i+1,n+1}, h_{i,n+1}\right) = 0 \tag{5.58}$$

$$\mathbf{M}_i \left(Q_{i+1,n+1}, Q_{i,n+1}, h_{i+1,n+1}, h_{i,n+1}\right) = 0 \tag{5.59}$$

where: **C** – equation derived from the continuity equation, **M** – the equation derived from the momentum equation.

Similar equations can be compiled for each point of the calculation scheme of the section for the time step $n+1$, which actually gives the above-mentioned system of equations. If the pipe is divided into m parts (m length steps Δx), we get $(2m+2)$ algebraic equations. This set of equations is solved simultaneously for the unknown variables h and Q for all calculation points at the time level $(n+1)\Delta t$, using the known values of the variables at the time level n and under the specified boundary conditions (Figure 5.9).

For N calculation points, there are $2N$ of unknown variables and it is possible to compile $2(N-2)$ equations. Thus, two boundary conditions are needed to obtain a solvable system of equations. These conditions must be formulated correctly, because it is not possible to enter a boundary condition in the form $Q(t)$ at the point where the height h is calculated as an unknown variable.

Implicit schemes are relatively more difficult to formulate and compile into a program, but if they are compiled and formulated correctly, they are very efficient and stable. The calculated time and length steps Δx and Δt can be chosen independently of each other. Implicit schemes are often evaluated as numerically unconditionally stable. This statement is based on linearised differential equations. However, for nonlinear differential equations, numerical stability is not ensured (it is not unconditional). Nevertheless, for sewer systems, the use of computational steps $\Delta t/\Delta x$ larger than required by Courant's criterion often does not cause any major problems.

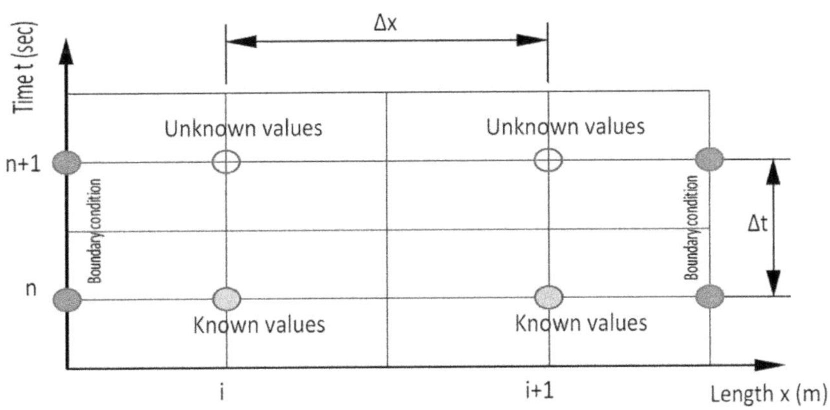

Figure 5.9 Concept of the implicit scheme.

5.4.3 Numerical stability, convergence

There are three basic aspects that need to be taken into account when selecting a suitable scheme when solving fluid flow modelling in sewer networks. They are:

1 stability of computation,
2 convergence,
3 computer capacity and speed requirements (computational time requirements).

Differential schemes can be stable or unstable. The differential scheme is stable when, as the differential network step (Δx or Δt) decreases, the solution of the problem comes (infinitely) closer to the solution of the solution of the task, defined by the corresponding boundary and initial conditions. This definition partially overlaps with the definition of convergence. The stability of the computational scheme is often checked by the so-called ε - scheme. In a task with a known numerical solution (for example, the values of the exact solution in the network nodes are known), an ε-error is made intentionally (instead of the $f_{i,j}$ value, the $f_{i,j}+\varepsilon$ value is taken) and its propagation is monitored. If the error in other nodes increases or remains the same, the given scheme is unstable with respect to the given equation and with the given network. Unstable schemes lead to numerical results that cannot be relied on, so they are not used.

These errors are compounded by a rounding error due to the computer's limited capabilities. This type of stability problems is related to numerical instability resulting from numerical operations. The differential scheme is stable if small numerical errors (arising, for example, when the rounding numbers in the computer's internal memory) do not increase and accumulate with a large number of arithmetic operations during the computation.

In principle, there is no universal criterion guaranteeing the stability of explicit schemes. Although it is true that in many cases the Courant criterion can be used, the depth and velocity of the flow vary with time and flow in the pipe, so it is not possible to precisely determine in advance the necessary time and length steps (Δx and Δt) for the computation. Because of this, almost all recent simulation programs use a variable timestep in their computational simulations, where the user typically defines the timestep range (minimal and maximal timestep). The necessary timestep Δt for the simulation is determined from Courant's criterion (Eq. 5.60) (Yen 1986), based on the most unfavourable results from the previous calculation step.

$$\frac{\Delta x}{\Delta t} \geq v + \sqrt{\frac{gA}{B}} \tag{5.60}$$

In addition, the selection of the time calculation step Δt is often a compromise of several criteria (Yen 1986).

The first criterion is the physical time required for the fluid stream to flow along the length of the calculation section. This criterion is not so important for a slowly changing steady flow (e.g., sanitary flows), and larger periods of time may be used in the numerical calculation. For rapidly changing flow (stormwater episodes), this criterion should be met to ensure the physical nature of the calculation. In sewers, where a small Δx is used in calculations compared to rivers and reservoirs, this criterion often requires Δt on the order of seconds, sometimes 1 or 2 seconds only.

Another criterion is the time interval of the used input data. Rain gauge or other input data (e.g., inflow hydrograms), the time step of which would correspond to the requirements for Δt according to the above-mentioned criteria (or Δt is greater than the requirements of the previous criteria), are often unavailable. The values of these input data must then be interpolated for the time step Δt used in the calculation.

The question of determining the convergence of the network method for a given task is also important. A convergent scheme is when the step x, Δt decreases, the approximate solution will be closer and closer to the exact solution, or whether for $\varepsilon>0$ such Δx, Δt is found that the error of the exact solution in the mesh with step Δx, Δt from the exact solution is less than ε. It is clear that for practical calculation, only stable and convergent methods can be used.

The considerations to be taken into account when selecting numerical schemes are the computer requirements of the individual schemes, including computer memory requirements, computer peripherals, computational time and costs, difficulties and problems in compiling the program, and finally the general use of the program. With the development of computers, the importance of many simplified methods has disappeared, e.g., kinematic wave and diffusion wave methods – at the current speed of computers, such simplifications are simply no longer necessary in terms of saving computational time. Therefore, it can be assumed that the current criteria for computer performance and capabilities will probably have to be changed in the near future.

When solving the Saint Venant equations, problems may arise originating from the assumptions made in deriving these equations. As mentioned above, these equations are applicable to slow-changing flow. The problem is especially the flow regime close to the critical flow, or the transition of sub-critical flow to supercritical flow regime and vice versa, or the transition between the free surface flow and pressure flow. Under these conditions, the equations describing the flow lead to instabilities in numerical solution, which can cause the computation crash.

The flow regimes are defined by the dimensionless Froude number Fr

$$\mathrm{Fr} = \frac{v}{\sqrt{gl}} = \sqrt{\frac{B}{gS}\frac{Q}{S}} \tag{5.61}$$

where: Fr – Froude number (-), v – average flow velocity (m/s), g – gravitational force (m/s^2), l – characteristic dimension (m), B – water level width (m), Q – flow rate (m^3/s), S – flow area (m^2). As the characteristic dimension can be used the average depth $h=S/B$ (m).

The momentum wave equation takes into account all the forces acting on the fluid. Nevertheless, problems arise in solving supercritical flow (Fr$>$1), because the complete Saint Venant equation is practically applicable only for the conditions of subcritical flow regime (Fr$<$1).

Critical motion appears in the case when the Froude number is equal to unity, which happens when (Hager 2010):

- at a specific constant flow rate Q=constant, the total stream energy head E is minimal,
- at a specific total energy head E=constant, the flow rate Q is maximal.

The critical motion, being a specific case of uniform flow (Fr=1) does not occur spontaneously under natural conditions. It is, however, important due to the fact that it constitutes the boundary between the area of tranquil flow, known also as subcritical motion (Fr$<$1) and the area of rapid flow – supercritical (Fr$>$1).

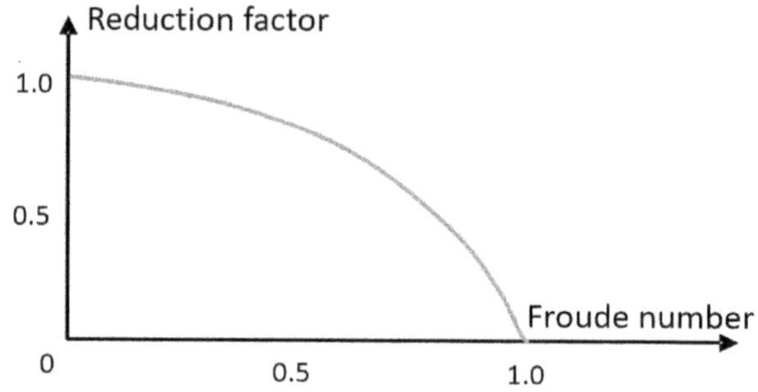

Figure 5.10 Reduction factor dependence on Froude number (flow regime).

In the case when the channel filling height is lower than the critical ($h<h_{kr}$) the subcritical - rapid flow takes place for which the Froude number is greater than unity. At the same time, the share of kinetic energy in the total energy increases, and that of potential energy decreases.

When the flow is tranquil ($Fr<1$), which is the case at a suitably small slope of the channel bottom, with a decrease of kinetic energy there is an increase of the potential energy of the sewage stream. Hence, the forces of gravity are the dominant forces in the system (Dąbrowski 2004).

For example, the MOUSE model eliminates the problems connected with the critical and supercritical flow ($Fr \geq 1$) by reducing Saint Venant equations to diffusion wave equations for the supercritical flow regime (DHI 1998). In the area of subcritical flow, the value of the terms $\partial Q/\partial t$ and $\partial/(Q^2/A)\partial x$ is reduced by a reduction factor, the value of which is chosen according to Figure 5.10.

5.4.4 *Initial and boundary conditions*

The unsteady flow in the sewer system depends on the initial depth and velocity (flow) in the drain at the beginning of the simulation, and also depends on the flow conditions at the inlets and outlets of the system. Mathematically, if the Saint Venant equations are used to solve unsteady problems, in addition to the initial condition, it is necessary to have two boundary conditions (input and output) to obtain a mathematically and logically correct solution describing the flow in the sewer network.

The initial conditions are the conditions in the pipe at the beginning of the simulation, i.e., at time $t=0$ it is necessary to define water depths in the initial sections of the sewer (i.e., $h(x,0)$) or water flow rates or flow rates ($v(x, 0)$, $Q(x, 0)$) in these points. In the stormwater sewer system, the initial depth and flow are theoretically zero; however, this sometimes causes problems in the simulation. Therefore, it is sometimes necessary to enter small values of flow rates as the initial flow, which usually corresponds to reality. In the case of combined sewer systems, this problem practically disappears, because the dry weather flow is usually considered as the initial flow.

Boundary conditions, when Saint Venant equations are applied to the inner section of a sewer network (i.e., not connected to the inlet or outlet of the system), are usually depths and flows from the end of the upstream section, which are considered to be identical with depths and flows at the beginning of the downstream section. Thus, the boundary conditions of the inner sections of the network do not have to be explicitly given, because they are actually implicitly given as values characterising the flow in neighbouring sections of the network. For "external" boundary conditions (i.e. for inflow and outflow from the sewer system), the upstream boundary condition is the inflow during the rain; these are usually the flow rates in the manhole or other inlet at the top of the section, i.e. $Q(0,t)$. If there is an outflow from the sewer network, i.e. in the part $x=L$, the boundary condition has to be defined. Typically, most often it is the time course of levels at the outlet from the sewer network. This boundary condition can also be specified implicitly, i.e. as the relationship between flow rate and depth (flow curve) or other relationships ($v=f(h)$, $h=f(Q)$, etc.).

Chapter 6

Runoff modelling in sewer networks

6.1 Reliability and safety as basic parameters of sewer network design

As indicated by design practice, guidelines, and literature considerations (Guan et al. 2015; Kirshen et al. 2015; Jiang et al. 2018; Jato-Espino et al. 2016), the basis for dimensioning a stormwater or combined sewer system is the maximum instantaneous catchment flow. To calculate its value, standards for the functioning of the sewer system should be established and defined as the capacity of the system to accept the predicted maximum stormwater flows with a frequency equal to the frequency of stormwater flooding (Table 6.1). The data in the table indicate that the design of the sewer system is based on the flooding occurrence, which can be defined as stormwater flooding leading to damage or disturbance of sewer operation (PN-EN 752: 2008).

As shown in Table 6.1, standard PN-EN 752 requires the frequency of events leading to stormwater flooding from sewer systems onto the surface. Hence, a relatively rare recurrence of such incidences is required: once in 10 years in rural areas, once in 20 years in residential areas, once in 30 years in city centres, industrial and service areas, and once in 50 years in underground communication facilities, passages, and crossings.

In accordance with standard PN-EN 752: 2008, to ensure these frequencies, the following modelled rainfall frequencies should be assumed in the sewer system design: once a year in rural areas and 2, 5, and 10 times a year for urban areas. However, due to the limited possibilities of modelling the flooding phenomenon (Brown et al. 2007; Hammond et al. 2015; Ke et al. 2020), the concept of surcharge, which indicates that the maximum computational reference level has been exceeded, was introduced to analyses verifying the operation of sewage systems (Table 6.2).

Table 6.1 Frequencies of modelled rainfall and allowable frequencies of overflow specified by PN-EN 752: 2008

Frequency of modelled rainfall [1 per C years]	*Category of area drainage standard*	*Allowable stormwater flooding frequency [1 per C years]*
1 per 1	Urban areas	1 per 10
1 per 2	Residential areas	1 per 20
1 per 5	City centres, industrial and service areas	1 per 30
1 per 10	Underground communication facilities, passages, and crossings	1 per 50

DOI: 10.1201/9781003546436-6

Table 6.2 Suggested frequencies of surcharge for verification calculations in accordance with PN-EN 752: 2008

Frequency of modelled rainfall [1 per C years]	*Category of area drainage standard*	*Frequency of surcharge events [1 per C years]*
1 per 1	Urban areas	2
1 per 2	Residential areas	3
1 per 5	City centres, industrial and service areas	Less frequently than 5
1 per 10	Underground communication facilities, passages, and crossings	Less frequently than 10

Besides the permissible overflow frequency assumed at the stage of sewer system design and the frequency of surcharge, standard PN-EN 752 provides guidelines on the methods that can be employed to determine the maximum outflow from the urban catchment. In the case of a catchment with an area below 200 ha or a time of concentration up to 15 minutes, provided that the rainfall intensity is constant, simplified approaches based on the rational method can be used to determine the outflow (chapter 2). Otherwise, or in the presence of a direct stormwater discharge via overflow to the receiving water, it is recommended that computer-aided simulations with implementation of kinematic and dynamic wave models should be performed.

The guidelines do not take into account the ongoing climate change (Huong & Pathirana 2013; Chawla et al. 2018; Hettiarachchi et al. 2018); hence, there is a need to define conditions of sewer system functioning necessitating modernisation of the sewer system. This issue is very important for ensuring a high standard of stormwater sewers. This problem was discussed by Siekmann & Pinnekamp (2011), who established the key parameters for the assessment of drainage system functioning:

- number of manholes with stormwater flooding:

$$N_j = \frac{\sum_{i=1}^{j} N_i}{N} \tag{6.1}$$

- stormwater flooding relative to the impervious unit area:

$$W_j = \frac{\sum_{i=1}^{j} W_i}{A} \tag{6.2}$$

where: A – impervious area in the catchment (m^2), j – number of manholes in the sewer system with stormwater flooding onto the ground surface (-), W_i – volume of stormwater flooding from the ith manhole in the analysed drainage system (m^3), N – number of manholes in the sewer system (-), N_i – number of overflowed manholes, W_j – specific flood volume per area (m), N_j – degree of flooding (-).

On the basis of these calculations, Siekmann & Pinnekamp found that no modernisation of the existing drainage system is recommended in the value range of $W_j = 0–7\,m^3/ha$, whereas it is necessary to reconstruct the sewer network in the range of $W_j > 15\,m^3/ha$ in order to reduce the volume of stormwater flooding. As regards the number of manholes with stormwater overflow, there is no need to modernise the existing land drainage system up to the value of $N_j = 0.17$. At $N_j > 0.42$, there is a considerable need to undertake measures for modernisation of the sewer network, i.e. construction of retention reservoirs and/or implementation of stormwater management systems (low impact development LID). The final decision rests with the managers of the sewer system.

This indicates that the formal and specific requirements for the design and assessment of the operation of sewer networks are now defined. These guidelines allow quantitative evaluation of the functioning of the drainage system, taking into account climate change. This is important, as it indicates the necessity of modernisation of sewer systems.

6.2 Fields of model application

As shown in the literature, the catchment outflow hydrograph is calculated with the use of physical numeric/mechanistic models, i.e., a model, based on the physical principles, represented by mathematical descriptions and simulated numerically. The surface runoff and stormwater flow through a channel network is described with the Saint Venant differential equation system (Eqs. 5.30 and 5.31). Other approaches include clustered parametric models, where the variability of physical and geographical characteristics of the catchment is averaged to one point, parametric models with distributed parameters, taking into account the variability of hydrological characteristics on the catchment surface. Since physical models take into account most of the phenomena occurring during surface runoff and sewer flow, they are the most accurate approaches. Nevertheless, given the problems in the determination of their parameters and their number, distributed parametric models are usually used in practical considerations. For calculations of surface runoff in these models, linear or non-linear models of reservoirs are often used for separated homogeneous catchment areas. These models of the water balance take into account not only rainfalls and runoff but also the phenomena of evaporation, infiltration, and groundwater inflow to the receiver. Regardless of the degree of its complexity, the model has to be calibrated with the use of rainfall and flow measurements to be a source of reliable results. Additionally, at the stage of model calibration, it is necessary to analyse the sensitivity of the models to eliminate negligible parameters and to carry out uncertainty analysis, showing the impact of the accuracy of model parameter estimation on simulation results. This is related to the large number of parameters of the surface runoff phenomenon and the flow in the sewer network included in the model. Given the complexity of the models for determination of surface runoff and the problems with the calibration thereof, the calculation of the inflow hydrograph can be carried out with a modified rational method, taking into account the distribution of rainfall intensity, the shape of the analysed catchment, and the mode of land use.

To sum up the chapter, it can be concluded that many methods have been proposed and are currently being used for the calculation of the catchment outflow hydrographs. When no rainfall and flow measurement results are available, the hydrograph should be determined with the physical, parametric, and statistical models described above, but their exact choice has not been clearly established. From a practical point of view, the only limitation of the approaches presented above is their computational complexity.

6.3 Process of modelling

This chapter explains the main steps of the overall process of the numerical modelling of flows in urban sewer systems. For better illustration of the data management, preparation as well as all other steps in the overall modelling process, it is demonstrated on the examples of the simulation program SWMM.

Hydrodynamic modelling of urban catchments is a complex task (Chen et al. 2012; 2018) requiring a number of steps that have a significant impact on the results of calculations and the reliability of the input data used for the simulation. Therefore, to obtain simulation results that will reflect real conditions prevailing during observed events, it is important to collect data that indicate the actual state. To achieve this goal, it is necessary to have adequate knowledge of the structure and operation of stormwater and combined sewer systems. Equally important is the synergy of various scientific disciplines providing spatial information and data (possible sources of acquisition and accuracy thereof), rainfall data (location and sensitivity of the rain gauge), data on measurements of sewer network filling and flows (canals, stormwater overflows, retention tanks, etc.), assumptions of model accuracy at the design stage, and discretisation of model subcatchments. Given the interactions between these factors, it seems advisable to develop a methodology for catchment modelling. In practice, this is a very difficult task, since access to such data is limited in some countries, while ample data are generally available in information systems and internet portals in many others. The problem has been recognised, and general principles of modelling have been developed (Mark et al. 2004; Kim et al. 2021). However, the guidelines do not include details on the possibility of acquisition of (spatial) data, determination of coefficients in subcatchments, and model calibration methods (Cristiano et al. 2017; Behrouz et al. 2020), which may be essential for the appropriate selection of measuring devices, their location, and time of measurements. Bearing in mind these remarks, an original methodology for construction of hydrodynamic catchment models has been proposed (Figure 6.1).

Spatial data are the basis for the development of a hydrodynamic urban catchment model. They are used for the estimation of the size of impervious area, size of subcatchments, land slopes, roughness coefficient values, and terrain retention capacity (Kleidorfer et al. 2009; Fletcher et al. 2013; Krebs et al. 2014). Nevertheless, it is crucial to establish the reliability of these data. The data used for the determination of catchment characteristics can be obtained from an orthophotomap, and the data from EGiBS can additionally be used to improve accuracy. They increase the accuracy of data on impervious areas (roofs, parking lots, roads, etc.). The accuracy of input data modelling based on a basic map should be regarded as a source of reference data. The data required for catchment modelling must also include information about the diameters, lengths, and slopes of sewer channels. The information can be obtained from the GESUT database or from the basic map.

The data mentioned above are incomplete and insufficient to construct a model of the channel network, as values of Manning's roughness coefficients, local resistance coefficients at the inlet and outlet of individual channels, and the hydraulically active cross-section of channel network are additionally required. These data can be obtained through inspection of the sewer network using appropriate mobile cameras (Hertzberg et al. 1998; Duran et al. 2002; Bin Ali et al. 2011). The results of such investigations have a large impact on the correct selection of the values of catchment model coefficients determining hydraulic conditions (Dirksen et al. 2013; Iooss & Lemaître 2015; Caradot et al. 2018). When no cameras can be used, Manning's roughness coefficient values are determined based on the type of material used for building the channel, its age, or literature data (Thorndahl 2009; Szeląg et al. 2016; Fatone et al. 2023).

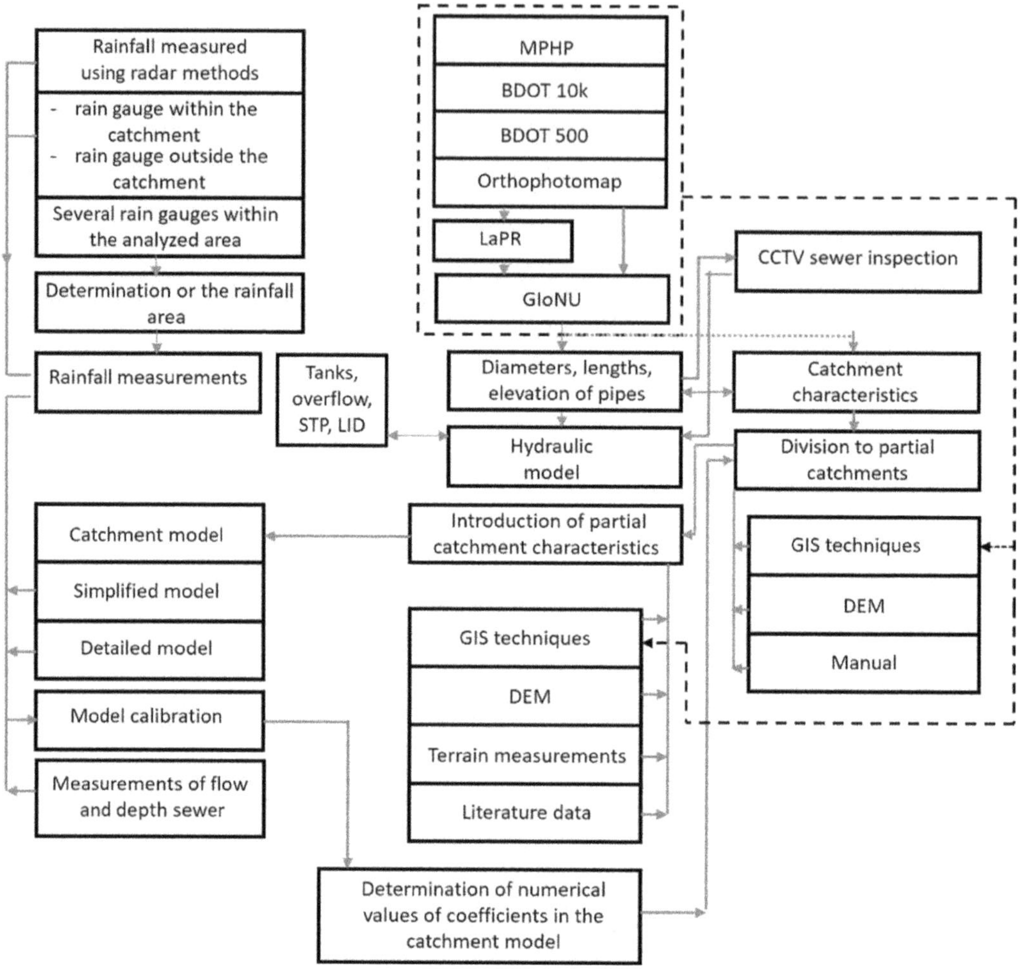

Figure 6.1 Conceptual algorithm for construction of a hydrodynamic catchment model.

At the step of preliminary calculations or in the unavailability of data on the sewer network, this information may be omitted at the stage of construction of the catchment model. This has a significant impact on the reliability of coefficient values determined while surface runoff is being modelled. To compensate for the omission of the canal retention value, the values of retention depth and terrain Manning's roughness coefficients will be higher than when the sewer network is included in the catchment model.

The next key stage in the development of the model is the division into subcatchments and the choice of methodology. It is possible to use concurrently the Digital Terrain Model (DTM), channel routes, and the course of linear drainage, which ensures the relatively smallest calculation error. In many cases, access to the DTM is limited, which may lead to potential errors and has an impact on the reliability and uncertainty of determined coefficients. In such cases, the basic map and the course of channels can be used, which may generate errors in the determination of catchment characteristics.

The selection of an appropriate method for analyses of the modelled catchment has an impact on the size of the subcatchment. This issue is highly important, as it has a significant effect on the variability of values of catchment surface area and calibrated coefficients. At this stage of modelling, it is also important to define the tasks to be fulfilled by the constructed catchment model, as it can be used for the estimation of catchment runoff, which requires less precise data in the model (Szeląg 2021b). It can also be used for the simulation of channel overflow and stormwater runoff, which requires highly detailed data. In practice, it seems correct to assume that the larger the modelled subcatchment area, the greater the uncertainty in the identification of the values of coefficients in the surface runoff model (Szeląg et al. 2016; 2020; Fatone et al. 2023). In turn, the lower the resolution of the catchment area is, the greater the variability of calibrated coefficient values may be. The accuracy of spatial data is highly important for the achievement of greater reliability of the determined catchment characteristics in a model based on detailed data than in a model with a limited number of subcatchments. At this stage, the use of DTM is essential to reduce the error in the assessment of the model coefficients if the sizes of the catchment area are optimally selected, i.e. the slope is not highly variable within the catchment and the land use is homogeneous.

An especially important aspect of model development is to fulfil the assumptions of the division into subcatchments. However, due to the diversity of data resources (sewer network, spatial data on land use, longitudinal catchment slope, etc.), the simplifications adopted at the stage of determination of the catchment resolution in the model have a considerable impact on the reliability of coefficients describing subcatchments, which influences the process of catchment model calibration. At the stage of collection of data for the development of a hydrodynamic model of the catchment, in terms of further calibration, field measurements are highly important for the reduction of the uncertainty of values of determined coefficients. The measurements may facilitate the determination of the values of calibrated parameters – coefficients in the Horton model, which may have a substantial impact on the convergence of calculation and measurement results and the reduction of relationships between the model coefficients. However, this is not a key issue at the stage of determination of model input data, since it is evident in many cases that the dynamics of the infiltration process have an inconsiderable effect on the results of calculations of the variability of the catchment outflow.

The data discussed above are necessary for the development of a hydrodynamic catchment model and the achievement of calculation results that are close to the actual course of rainfall events. The model must be calibrated to improve its predictive capabilities and ensure agreement between calculation and measurement results. The calibration requires gathering data on the variability of the flow in the sewer network with an appropriate resolution (Mancipe-Munoz et al. 2014; Obermayer et al. 2010; Mrowiec 2009). When measuring devices are unavailable, methods used in hydrology can be employed (Mrowiec 2009; Geberemariam 2015). One of the simplest devices that can be used is the stormwater level gauge, which can be made in the sewer network simply by marking the water level on the walls of channels, straight-through and connection manholes, and retention reservoirs with a washable marker. Obviously, this only allows observation of the maximum stormwater depth in a given rainfall event. Nevertheless, such measurements conducted in a long sewer network and at many points can be a valuable source of useful information for the calibration of the catchment model. The coefficient values obtained in this way during model calibration will certainly be burdened with much greater uncertainty than in the case of a model calibrated based on continuous flow or filling measurements. However, in a large number of cases, this is the only possible solution in the absence of measuring tools, which is usually associated with financial, organisational, or legal problems. The flow measurement method largely determines the type of model calibration, i.e. continuous based on time series or discontinuous based on separate rainfall events. Certainly, the model calibration based on continuous time series provides greater reliability of the coefficients identified in the model. Due to technical problems, there may be issues with collecting continuous data at the flow measurement stage, which result from the specificity of the operation of the sewer network. In addition to flow measurements, rainfall measurements are essential for the calibration of the model. In this case, the resolution of rainfall data, the location of the rain gauge, and the type of the measuring device are highly important. The catchment size and the number of rain gauge stations located within the catchment may significantly hinder the achievement of reliable values of model coefficients in this case. In large urban catchments, the rainfall field distribution may vary; hence, it should be taken into account at the stage of model calibration, as it has a considerable impact on the reliability of the model coefficients and their uncertainty. This issue is of great importance at the stage of constructing models of large sewer networks. Therefore, it is necessary that the number of rain gauges in the catchment area should ensure high convergence of calculation and measurement results. Importantly, the number of rain gauges should not be too high, as their operation generates significant costs. Therefore, the number of rain gauges in the catchment should be optimised taking into account local conditions.

To sum up these considerations, it can be concluded that catchment modelling is a complex task requiring the collection of spatial data on catchment characteristics related to the sewer network and conditions prevailing therein (flows and filling in the channels), as well as rainfall data with their variability in the analysed area. Collection of highly reliable data is not easy, and the identification of data that have the greatest impact on simulation results is complex. This is a result of numerous interactions between the analysed model variables, as confirmed in the model calculation scheme (Figure 6.1). It is obvious that, in the case of higher accuracy of measurements of the rainfall field in the catchment and less detailed information about sewer channels and conditions prevailing therein, there is a strong non-linear interaction between the considered variables, and the determination of its impact on the reliability of coefficients identified in the model is highly complicated. There are many such interactions at the model development stage, and the analysis thereof is a very complex task requiring the implementation of advanced statistical models. This is currently the subject of numerous studies in this scope, but it is difficult to give an unambiguous answer to the question of how to design a model, which data types ensure a relatively small error in the identification of model coefficients, and which data may result in a substantially greater error. This is associated with the fact that, despite the availability of many data sources and methods, not all data are typically quantitative, and some data are qualitative. Hence, there is a need to standardise procedures and distinguish groups of data clearly by the introduction of quality classes to obtain reliable values of coefficients calibrated in the model.

The chapters below present the stages of construction of a model in the SWMM program, with particular emphasis on the simulation of surface runoff, retention reservoirs, storm overflows, pump systems, and rainfall data.

6.4 Model calibration and verification

Besides gathering reliable data for the generation of the hydrodynamic catchment model, its correct operation, and the stability of numerical solutions, model calibration is of fundamental importance (Mrowiec 2009; Price & Vojinovic 2011). It aims at the accurate determination of the values of coefficients (describing partial phenomena) to achieve the best fit of the calculation results and field measurements. This is a key issue, as a calibrated catchment model is a reliable source of knowledge of phenomena occurring in analysed sewer networks (Szeląg et al. 2016; Kiczko et al. 2018). Model calibration is a complicated task, especially in the case of urban catchment models, where the modelled phenomena are non-linear and there are interactions between identified coefficients; this makes the correct determination thereof difficult. Another factor that has a substantial impact on the calibration task is the number of coefficients included in the task. In many cases, a number of model coefficients turn out to have a negligible effect on calculation results, which implies that they can be eliminated at the calibration stage and their values provided in the literature can be assumed (Wang 1991; Bates & Campbell 2001). This task is referred to as the sensitivity analysis of the model (Barco et al. 2008; Bruni et al. 2015). It is useful for the simplification of the model calibration stage and is often used in engineering practice.

6.4.1 Measures of the goodness-of-fit of calculation and measurements results

At the stage of development of the concept of a model, it is important to define the type of variables to be predicted and the resolution with which they are to be saved. This is an important issue, as the model is calibrated based on these variables, which influences the reliability of model coefficients and calculation results. As shown by the literature review (Price & Vojinovic 2011), the following types of data are the basis for hydrodynamic model calibration:

- continuous (stormwater volume and quality, channel filling),
- binary (pressure flow, stormwater discharge via overflow, tank overflow, first-flush phenomenon, etc.).

In the case of continuous data, the calibration may include various parameters in the observed variability of processes. These can be extreme values (e.g. maximum instantaneous flow, maximum channel filling), total outflow hydrograph volume, or variability of discrete $Q(t)$ or $S(t)$ values resulting from a rainfall event. In the case of binary data, only the occurrence of the analysed phenomenon is important. In practice, these variables are much easier to define and determine in terms of stormwater volume than in the case of continuous data. In analyses of binary phenomena, it is sufficient to state that they have occurred, without the need to specify their exact numerical value (Van Bijnen et al. 2012; Guan et al. 2015).

Bearing these considerations in mind, the following goodness-of-fit measures are used at the model calibration stage:

continuous data:

$$\text{correlation coefficient}(R): R = \frac{\sum_{i=1}^{N}\left(y_{\text{obs}} - \overline{y_{\text{obs}}}\right)\cdot\left(y_{\text{obl}} - \overline{y_{\text{obl}}}\right)}{\sqrt{\sum_{i=1}^{N}\left(y_{\text{obs}} - \overline{y_{\text{obs}}}\right)^2}\cdot\sqrt{\sum_{i=1}^{N}\left(y_{\text{obl}} - \overline{y_{\text{obl}}}\right)^2}} \tag{6.3}$$

$$\text{modified correlation coefficient}\left(R_z\right): R_z = \frac{\sum_{i=1}^{N}\left(y_{\text{obl}} - \overline{y_{\text{obs}}}\right)^2}{\sum_{i=1}^{N}\left(y_{\text{obs}} - \overline{y_{\text{obs}}}\right)^2} \tag{6.4}$$

$$\text{Nash – Sutcliffe coefficient}\left(\text{NS}\right): \text{NS} = \frac{\sum_{i=1}^{N}\left(y_{\text{obs}} - \overline{y_{\text{obs}}}\right)^2 - \sum_{i=1}^{N}\left(y_{\text{obs}} - y_{\text{obl}}\right)^2}{\sum_{i=1}^{N}\left(y_{\text{obs}} - \overline{y_{\text{obs}}}\right)^2} \tag{6.5}$$

$$\text{relative error}\left(\text{BW}\right): \text{BW} = \frac{\sum_{i=1}^{N}\left(y_{\text{obl}} - y_{\text{obs}}\right)}{\sum_{i=1}^{N} y_{\text{obs}}} \tag{6.6}$$

$$\text{mean relative error}\left(\text{MAPE}\right): \text{MAPE} = \frac{1}{N}\sum_{i=1}^{N}\left|\frac{y_{\text{obs}} - y_{\text{obl}}}{y_{\text{obs}}}\right| \cdot 100\% \tag{6.7}$$

$$\text{mean absolute error}\left(\text{MAE}\right): \text{MAE} = \frac{1}{N}\sum_{i=1}^{N}\left|y_{\text{obs}} - y_{\text{obl}}\right| \cdot 100\% \tag{6.8}$$

$$\text{Root mean square error}\left(\text{RMSE}\right): \text{RMSE} = \sqrt{\frac{1}{N} \cdot \sum_{i=1}^{N}\left(y_{\text{obs}} - y_{\text{obl}}\right)^2} \tag{6.9}$$

$$\text{quotient of observed maximum to calculated maximum}\left(\delta\right): \delta = \frac{y_{\text{obs,max}}}{y_{\text{obl,max}}} \tag{6.10}$$

where: y_{obs} – observation results, y_{obl} – measurement results, N – number of measurements.
binary data:

$$\text{sensitivity}\ \left(\text{SENS}\right): \text{SENS} = 100 \cdot \frac{\text{TP}}{\text{TP} + \text{FN}} \tag{6.11}$$

$$\text{specificity}\ \left(\text{SPEC}\right): \text{SPEC} = 100 \cdot \frac{\text{TN}}{\text{FP} + \text{FN}} \tag{6.12}$$

$$\text{calculation error}\ \left(R_z^2\right): R_z^2 = 100 \cdot \frac{\text{TN} + \text{TP}}{\text{TP} + \text{TN} + \text{FP} + \text{FN}} \tag{6.13}$$

Table 6.3 Example of a classification table

Classification		*Predicted decisions*	
		positive	*negative*
Observed decisions	Positive	True positive (TP)	False negative (FN)
	Negative	False positive (FP)	True negative (TN)

where: TN, TP, FP, FN – number of indications classified in accordance with Table 6.3 as true negative, true positive, false positive, false negative, respectively.

In addition to these measures of the goodness-of-fit of the calculation and measurement results (continuous data), especially in the case of statistical models, other measures can be used to assess the accuracy of the generated model:

$$\text{maximum mean percentage error}\left(\text{PE}\left(\text{min}\right)\right) \quad \text{PE}\left(\text{max}\right) = \max\left\{\frac{\left(\varepsilon_{Vt} - \varepsilon_{Vn}\right)}{\varepsilon_{Vt}} \cdot 100\%\right\} \tag{6.14}$$

$$\text{minimum mean percentage error}\left(\text{PE}\left(\text{max}\right)\right) \quad \text{PE}\left(\text{min}\right) = \min\left\{\frac{\left(\varepsilon_{Vt} - \varepsilon_{Vn}\right)}{\varepsilon_{Vt}} \cdot 100\%\right\} \tag{6.15}$$

To obtain a model with the best predictive abilities, the following is advisable: (MAE, MAPE, RMSE, PEmin, and max) values → min, while the R value → max. In the case of classification models for simulation of binary data, the following should be achieved: (SEN, SPEC, R_z^2) values → max, i.e. up to 100%.

Additionally, to assess the model prediction accuracy, the so-called information criteria can be used for the evaluation of the statistical correctness of the goodness-of-fit of empirical data, with consideration of the complexity of the model associated with the number of estimated parameters. This can be achieved with the use of the Akaike Information Criterion (AIC) expressed by the following equation:

$$\text{AIC} = \left[2.83788771 + \ln\left(\frac{\text{SSE}}{N}\right)\right] \cdot N + 2L_p \tag{6.16}$$

where: L_p – number of estimated parameters, SSE – sum of squared errors.

6.4.2 *Deterministic (manual) calibration*

In the process of generation of a reliable model for the simulation of the operation of the sewer network, the sensitivity analysis is followed by the stage of calibration of model coefficient values that have a crucial impact on the goodness-of-fit of calculation and measurement results. Calibration with the manual method (or successive iterations) is a tedious task. Manual calibration can be performed in two different ways:

- discontinuous, i.e. based on separate rainfall-runoff events in time series,
- continuous, i.e. modelling of a time series of several rainfall events.

Table 6.4 Values of the goodness-of-fit of models (R_z, NS, BW) and measurement results

R_z	*NS, BW*	*Model fit*
< 0.05	> 0.85	Excellent
0.05–0.10	0.65–0.85	Very good
0.10–0.20	0.50–0.65	Good
0.20–0.40	0.20–0.50	Weak
> 0.40	0.20	Unsatisfactory

The process of catchment model calibration varies and depends on rainfall events (time series of $Q(t)$ and $P(t)$ values) selected to identify model coefficients and subjected to verification of their values in independent events. In the first stage of calibration, the determination of coefficients in subcatchments is based on spatial data (orthophotomap, basic map, satellite photos, etc.) or literature reports (Szeląg et al. 2022b; Caradot et al. 2020). On this basis, runoff hydrographs or the variability of $h(t)$, $S(t)$, or other dependent variables are calculated for selected rainfall-runoff events (time series). The results of these analyses are used for the determination of the goodness-of-fit values and comparison thereof with theoretical values. In the case of high convergence of calculation and measurement values (Table 6.4), the rainfall-runoff model is verified in independent events or in time series $Q(t)$. When the determined goodness-of-fit values are not satisfactory, it is advisable to correct the assumed values of model coefficients. This task has a discrete character and depends on many factors. At this stage, it is important to make an appropriate choice of the value of coefficient(s) to be modified. Currently, there is no clear-cut methodology, but the selection of coefficient(s) to be corrected can be based on sensitivity analysis results.

At this stage of model calibration, the corrected coefficients should be selected properly to achieve the best fit of calculation and measurement results.

The parameter calibration process in the catchment model can be modified by the identification of parameter values with optimisation methods (genetic algorithms, ant colony, bee colony, cuckoo egg, Levenberg-Marquardt, Powell, and Nadler-Mead). In these models, it is advisable to "manually" provide determined coefficient values as the initial values to be calibrated. This approach is confirmed in many studies on the analysis of the operation of engineering and technical facilities and devices (Mrowiec 2009; Del Giudice & Padulano 2016; Pang et al. 2020). An alternative solution for the calibration of models and the identification of coefficients is the use of parametric machine learning methods (artificial neural networks, support vectors, and regression tree methods with modifications, etc.).

6.5 Model sensitivity analysis

It is advisable to carry out a sensitivity analysis before the calibration of the generated hydrodynamic catchment model (Holman-Dodds et al. 2003; Geberemariam 2015; Ballinas-González et al. 2016). The sensitivity analysis is an important tool for the analysis of mathematical models, as it facilitates the determination of the degree of change in the response of the model to changes in input parameters and the identification of the strength of the relationship between the uncertain input and output. This is highly important for the potential elimination of variables that have a low impact on the model response. The choice of the sensitivity analysis method depends on the sensitivity measure used and the accuracy of estimation of the sensitivity measure and computational costs. On the basis of the choice of the sensitivity measure and the variability of model parameters, the sensitivity analysis methods can be classified as presented below.

Local sensitivity analysis (estimation of model sensitivity with reference to the variability of parameters and determination of gradients and derivatives at points). When the dependent and independent variables have identical dimensions, only partial derivatives are determined as follows:

$$y_{x_i} = \frac{\partial y}{\partial x_i} \tag{6.17}$$

To compare variables with different dimensions, the derivatives should be normalised and the sensitivity function should be determined as follows:

$$S_{x_i} = \frac{\partial y}{\partial x_i} \cdot \frac{\sigma_{x_i}}{\sigma_{y_i}} \tag{6.18}$$

Equation 6.18 is a function of variables x_i, which is the product of the standard deviation σ_{x_i} and the partial derivative of the original function $\partial y/\partial x_i$. The number of sensitivity functions that can be generated for one primitive function is equal to the number of variables (and parameters) included therein. Since the determination of local derivatives is not complicated, this method is widely used at the stage of generation of models for the prediction of the quantity and quality of stormwater or sanitary wastewater (Skotnicki & Sowiński 2009; Mancipe-Munoz et al. 2014). Nevertheless, the sensitivity coefficients obtained in this way have a local nature, and the analysed impact of variability Δx_i at point x_i does not exceed 10% (Barco et al. 2008; Mrowiec 2009). When the standard deviation (σ_{x_i}) is unknown, values σ_{y_i} and Eq. 6.18 can be expressed as follows:

$$S_{x_i} = \frac{\partial y}{\partial x_i} \cdot \frac{x_i}{y(x_i)} \tag{6.19}$$

Global sensitivity analysis (testing the behaviour of the system over a full range of parameters) facilitates assessment of the impact of selected independent model variables on the simulation results in a wide range of variability. One of the frequently employed approaches in global sensitivity analysis is the SRC (standard regression coefficient) method, which is commonly used for analysis of hydraulic and hydrological models (Szeląg et al. 2020; Fatone et al. 2021). In this approach, the values of the dependent variable (model outputs, i.e. maximum flow, outflow volume, overflow volume, etc.) in the SRC method are determined with the following equation:

$$y = a_1 \cdot x_1 + a_2 \cdot x_2 + a_1 \cdot x_3 + \cdots + a_j \cdot x_j \tag{6.20}$$

where: a_1, a_2, a_3, a_j – empirical coefficients determined with the maximum likelihood method, x_1, x_2, x_3, x_j – standardised independent variables expressed as:

$$x_j^* = \frac{x_j - \min\{x_j\}}{\max\{x_j\} - \min\{x_j\}} \qquad x_j^* \in [0;1] \tag{6.21}$$

In the model described by Eq. 6.20, the calculated values of coefficients α_j allow the determination of the impact and ranking of the analysed predictors (x_j) on the modelled dependent variable (Szeląg et al. 2016; 2020; 2022c; 2023). In many cases, at the stage of analysis of complex models describing catchment runoff or stormwater overflow, it is necessary to perform simulation with the use of the tested model, assuming a random distribution of calibrated parameters (e.g. depth of storage depression of impervious and pervious areas, Manning's roughness coefficient of impervious areas, channel roughness coefficient, etc.) to obtain Eq. 6.21. The Monte Carlo method is suitable for this purpose. It is applied to solve problems in which random numbers with a certain distribution are used as samples describing the process. Therefore, this method is based on numerical calculations, where the process is stochastic and the numerical solution is replaced by a task from the probability domain. The Monte Carlo method is based on multiple sampling of the variable X value (j – identified hydrodynamic model parameters) in a series of independent samples. With a sufficiently large number of simulations, this method yields a result expressed as the arithmetic mean of modelled values. Calibrated parameters determined with the Monte Carlo method can be used in the hydrodynamic model, and hydrological parameters (outflow volume, stormwater quality, stormwater flooding, etc.) can thus be calculated. These results are used as input data to construct the model described by Eq. 6.20. At the stage of analyses, models differing from Eq. 6.20 are used as well, since the relationship between the modelled dependent variable and independent variables is non-linear in many cases (Cosenza et al. 2014; Szeląg et al. 2020). In these cases, Eq. 6.20 includes the effect of interactions between independent variables (Szeląg et al. 2016). There are also highly non-linear models facilitating identification of the values of coefficients α_j with appropriate interpretation (Szeląg et al. 2022b).

6.5.1 *Probabilistic calibration (GLUE+GSA)*

The basis of the GSA-GLUE technique is the probabilistic formulation of the parameter identification task with the GLUE technique. The most frequently used approach consists in searching for one established set of parameters, which ensures the best fit of simulation results obtained in the model to measured values. The probabilistic approach is associated with the observation that, in the case of models with a large number of parameters, an inverse solution targeted at the determination of parameter values from measurement data may lead to many equivalent results (Beven & Binley 1992). The various combinations of parameters ensuring satisfactory representation of the dynamics of analysed processes by the model result from the lack of sufficient data for the unambiguous determination of the parameters. One of the possible solutions to this problem is to assume the probabilistic form of the parameter identification task. In such a case, the existence of one optimal combination is rejected in favour of a random parameter distribution, the properties of which are determined in the identification process. The probabilistic description of the parameter space in GLUE is also used for the description of the uncertainty of the model itself, where the variability of parameters reflects not only the insufficient amount of data but also the imperfection of representation of the actual process (inaccuracy of data, inadequacy of mathematical description, introduced simplifications) (Romanowicz & Beven 2003). The problem of identification of parameters in the GLUE method is formulated according to the Bayesian estimation as:

$$M(X \mid Q) = \frac{M(X)\,M(Q \setminus X)}{M(Q)} \tag{6.22}$$

where: X – vector of model parameters, $M(X)$ – a priori distribution of parameters, Q – vector of measured flow values, $M(Q\backslash X)$ – likelihood function, $M(X\backslash Q)$ – resultant a posteriori distribution, $M(Q)$ – scaling element defined in such a way that $\sum M(X\backslash Q)=1$.

The *a priori* distribution $M(X)$ reflects the initial assumption about the variability of parameters. In the case of mathematical models describing surface runoff, there is usually no information about this distribution, and only the range of permissible values of parameters associated with their physical interpretation may be known. In the case analysed in the study, a uniform distribution was assumed, which is a common practice in this type of applications (Romanowicz & Beven 2003; Beven 2001). The identification of model parameters in the GLUE method consists in transformation of *a priori* into *a posteriori* distribution using the likelihood function, which determines the probability of a combination of parameters depending on the goodness-of-fit of calculation and measurement results. The following considerations are based on the likelihood function expressed by the formula below (Romanowicz & Beven 2006; Dąbkowski et al. 2010):

$$M(Q\backslash X)=\exp\left[\frac{-r}{k\cdot V(r)}\right] \tag{6.23}$$

where: k – scaling parameter, $V(\cdot)$ – variance, r – mean of the sum of the square deviation of the simulated value from the measured value calculated as follows:

$$r=\frac{1}{N}\sum_{k=1}^{N}\left(Q_{m,k}-Q_{s,k}\right)^2 \tag{6.24}$$

where: $Q_{m,k}$ – measured rainwater flow intensity in the cross-section closing the catchment with a 1-minute time step k, $Q_{s,k}$ – simulated rainwater flow intensity in the SWMM program.

In accordance with the GLUE methodology, Eq. 6.23 is solved with the Monte Carlo method. In the first step, a sample of parameters from the assumed *a priori* distribution is created. For each combination of parameters constituting vector $X=[x_1, x_2, \ldots, x_i]$, the model is run and, based on calculated and measured outflow Q hydrographs, the likelihood function values and *a posteriori* distributions are determined.

The Global Sensitivity Analysis variance method described by Archer et al. (1997) is an extension of the Generalised Likelihood Uncertainty Estimation (GLUE) approach. It can be used independently and complement the GLUE methodology. The sensitivity analysis allows the determination of the impact of parameter variance on the model output variance. This facilitates the identification of parameters that have a significant impact on the uncertainty of the entire model. The Global Sensitivity Analysis represents a group of variance methods based on the decomposition of variance (Saltelli et al. 2007):

$$V(Y)=V\left[E\left(Y\mid X_i=x_i^*\right)\right]+E\left[V\left(Y\mid X_i=x_i^*\right)\right] \tag{6.25}$$

where: Y – model output, $E(\cdot)$ expected value, $E\left(Y\mid X_i=x_i^*\right)$ and $V\left(Y\mid X_i=x_i^*\right)$ are the expected value and model output variance, respectively, for variable parameters X at a value of the i-th parameter X_i set at x_i^*. The additional functions $E(\cdot)$ and $V(\cdot)$ encompassing these two elements are defined in relation to parameter X_i assuming different values in its variability range. When the parameter to which the model is highly sensitive has been determined, the output variance $E\left[V\left(Y \mid X_i = x_i^*\right)\right]$ expected for different values X_i should be relatively small, as term $V\left(Y \mid X_i = x_i^*\right)$ does not include the variability of this parameter. The significance of the parameter will be reflected in a large variance of term $V\left[E\left(Y \mid X_i = x_i^*\right)\right]$, when the expected output value changes with parameter X_i.

The model output variance described by the second term of Eq. 6.25 $V\left[E\left(Y \mid X_i = x_i^*\right)\right]$ is related to the direct impact of the variance of parameter X_i. On this basis, the GSA method proposes a measure of the sensitivity of the model to the i-th parameter, i.e. the so-called first-order sensitivity index S_i (Archer et al. 1997):

$$S_i = \frac{V\left[E\left(Y \mid X_i = x_i^*\right)\right]}{V\left(Y\right)} \tag{6.26}$$

The decomposition of variance described in Eq. 6.26 can also be performed by determination of values of all parameters ($X_{-i} = x_{-i}^*$) other than X_i, which yields:

$$V\left(Y\right) = V\left[E\left(Y \mid X_{-i} = x_{-i}^*\right)\right] + E\left[V\left(Y \mid X_{-i} = x_{-i}^*\right)\right] \tag{6.27}$$

As in the case of Eq. 6.25, the first term in Eq. 6.27 $V\left[E\left(Y \mid X_{-i} = x_{-i}^*\right)\right]$ describes a part of model variance resulting from the direct interaction of parameters X_{-i} with output Y. The second term describes the share of parameter X_i in interaction with the other parameters X_{-i}. On this basis, the total sensitivity index (S_{Ti}) is defined, which describes the overall sensitivity of the model to parameter X_i calculated as follows (Archer et al. 1997):

$$S_{Ti} = \frac{E\left[V\left(Y \mid X_{-i} = x_{-i}^*\right)\right]}{V\left(Y\right)} \tag{6.28}$$

The major problem in determining the values of sensitivity indices in Eqs. 6.26 and 6.28 is the necessity to know the variance of the parameters. One of the possible solutions is to combine the sensitivity analysis with the identification of parameters using the GSA-GLUE method (Beven & Binley, 1992). The sensitivity analysis is performed with reference to an index that has the same form as the likelihood function used in GLUE. This allows the analysis of the model variance identified as uncertainty in the GLUE method. The decomposition of variance is performed based on the Monte Carlo technique, where it is necessary to maintain an appropriate sample structure due to the need for determination of individual values of groups of parameters ($X_i = x_i^*$, $X_{-i} = x_{-i}^*$) in accordance with Eqs. 6.26 and 6.28. Most frequently, an appropriately constructed sample is used both for the identification of parameters with the GLUE method and for sensitivity analysis. The calculation scheme shown in Figure 6.2 presents the stages of GSA-GLUE analyses.

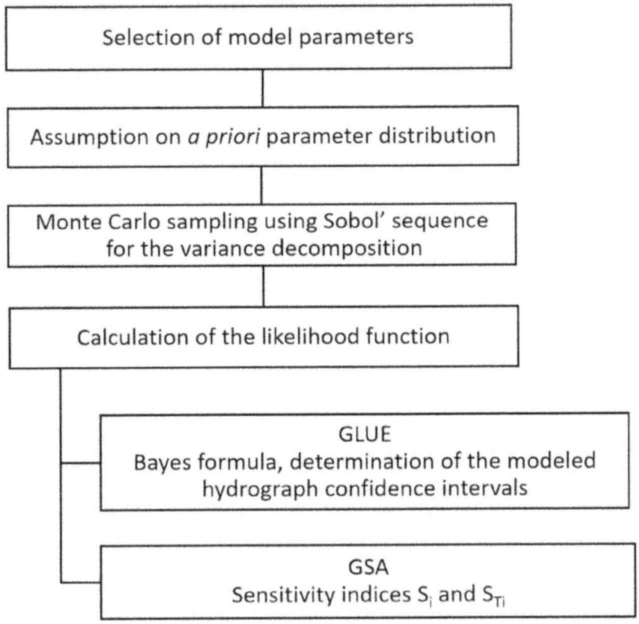

Figure 6.2 Calculation scheme in the GLUE + GSA method (Szeląg et al. 2016).

The methodology described above is quite frequently employed at the stage of calibration of urban catchment models, which is confirmed by the large number of studies in this field (Muleta & Nicklow 2005; Song et al. 2015). These papers show a significant impact of the uncertainties identified in the coefficient models on the results of the simulation of stormwater quantity and quality (Dotto et al. 2012). Nevertheless, this aspect is currently not taken into account in methods for the assessment of the function of sewer networks or for the dimensioning of their facilities.

On the basis of these considerations, it can be assumed that the construction of a hydrodynamic catchment model characterised by high convergence of calculation and measurement results and reliable values of calibrated model coefficients is not comparable and does not necessarily always mean the same. On the one hand, it is possible to achieve high convergence of measurement and calculation data of the measured hydrographs. However, a question arises as to whether this relationship is also maintained for other rainfall events. An unambiguous answer to the question is very difficult, as it also depends on the procedure of catchment model calibration. The inaccuracies of the designed hydrodynamic catchment model are highlighted by the results of calibration carried out with the probabilistic method (Szeląg et al. 2016; 2022c). However, if the model input data are additionally burdened with uncertainty and measurement errors, the optimal calculated values of model coefficients may be unreliable, and further analyses may thus be biased. In practice, this implies that a model for analysis of sewer network operation that ignores the uncertainty of some coefficients increases doubts and may be a source of calculation errors and problems at the stage of operation of the designed and modernised sewer network. Therefore, it is advisable to search for innovative calibration methods that take into account both the complexity of catchment models and the uncertainty of input data used for the construction of the model.

Reliable results of calculations of catchment outflow variability are important, as they are the basis for the assessment of the operation of a sewer network and its facilities. As indicated by literature data (Kiczko et al. 2018; Szeląg et al. 2019) and the authors' experience, retention tanks play a key role in the reduction and control of the wastewater stream at the catchment outflow.

6.6 Overview of the most used urban rainfall – runoff models

6.6.1 *SWMM (Storm Water Management Model)*

SWMM is a program developed by EPA (US Environmental Protection Agency). Its first version was released in 1971, but the program has been continuously expanded and modified to date. It is used for modelling stormwater sewer networks as well as modelling and designing combined and sanitary sewer systems, including pressure sewers. The SWMM program facilitates modelling both the variability of the quantity and quality of stormwater in sewer networks. As emphasised by many researchers (Rossman 2010; Sañudo et al. 2020), although the SWMM program helps to model the phenomenon of stormwater or sewage flooding, determination of its extent requires the implementation of more complex methods and integration with other tools/programs (GIS for spatial analyses, MATLAB, PYTHON, etc.), which are equipped with computational algorithms for modelling the overflow layer depth and extent. Although the SWMM program is primarily intended for modelling urban catchments, it is also used for simulation of the quantity and quality of runoff in agricultural and river catchments (Dong et al. 2022; Chen et al. 2013). It is also possible to integrate catchment and watercourse models, as well as model the quantity and quality of stormwater along their length. This is a significant advantage of the program, as it can be used to study the impact of changes introduced within the catchment area on the hydraulic conditions in the receiver (Kubrak 1998). It arouses considerable interest, as confirmed by a large number of studies in this field (Guo et al. 2012; Yang et al. 2022).

The SWMM program consists of several modules:

RAIN – allows entering rainfall data as rainfall intensity, rainfall depth, and cumulative rainfall curves; noteworthy, the structure of the program and files with the calculation results saved in the xxx.inp. format can be modified to the txt. Format, and multiyear rainfall series can be introduced;

TEMPERATURE – allows entering and collecting temperature data related to snow and evaporation and melting, evaporation of water from surface;

RUNOFF – used to model surface runoff; what is noteworthy, is that the program offers a possibility of defining variability of stormwater entering the sewer network as a function of time or modelling its quantity;

EXTRAN – used for hydraulic calculations of the sewer network; Saint Venant equations are used to describe the water table in the system; it is also possible to simplify this calculation model;

STORAGE – allows simulation of the operation of sewer network objects, which enable the retention and delay of stormwater outflow from the catchment;

LID (Low Impact Development) – used for designing sustainable rainwater management systems;

QUALITY – used for modelling the quality of stormwater flowing out of the catchment;

STATISTICS – facilitates statistical analysis of results.

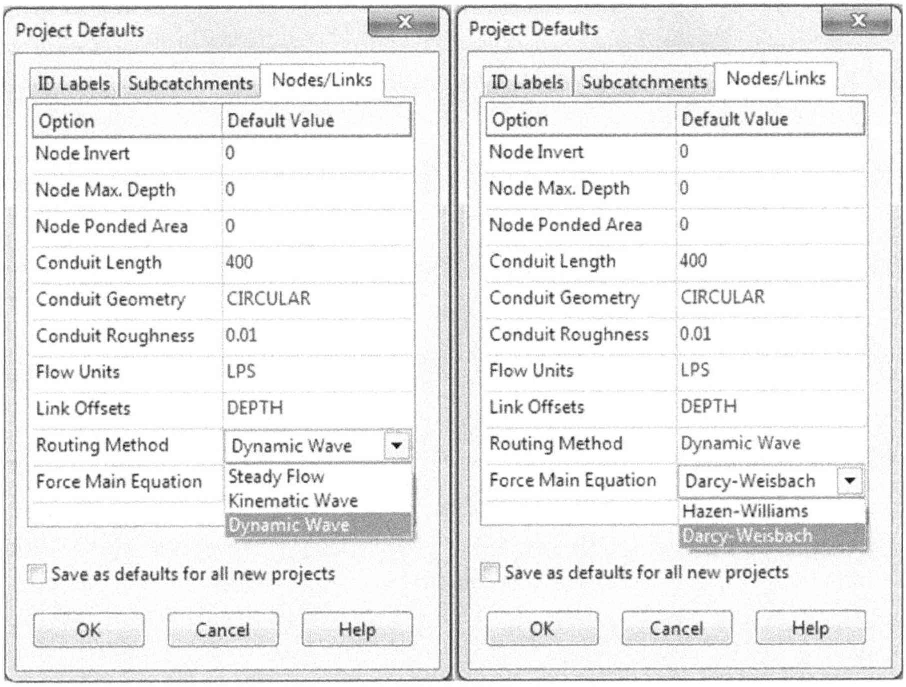

Figure 6.3 Dialog box for definition of metric units of results and models for calculation of the stormwater table system in channels.

6.6.2 *Data preparation and model arrangement*

Before starting to construct a catchment model in SWMM, it is necessary to define the system of e.g., metric units. The user can choose from other systems than SI; however, after setting Project/Defaults/Node/Links and selecting Flow Units: LPS, the data describing catchments and channels are entered in the SI metric system (Figure 6.3). Additionally, at this stage, it is advisable to determine the method for calculation of the stormwater depth in the channels (Dynamic/Kinematic/Steady) and to establish the relationship (Force Main Equation) used for determination of pressure losses in the pipes during the pressure flow. Another important element at the stage of generation of the model is to establish the time step (Routing) at which the Saint Venant equation system is solved, and the mode of recording calculation results (Reporting) depending on the period covered by the calculations (Figure 6.4).

6.6.3 *Subcatchment*

At the stage of creation of the hydrodynamic model, after previous determination of the model that will be used to simulate the stormwater depth in the channels, there is a need to identify (assign) the characteristics of subcatchments. When the calculations relate to the simulation of stormwater network operation, it is necessary to determine the relevant hydrological

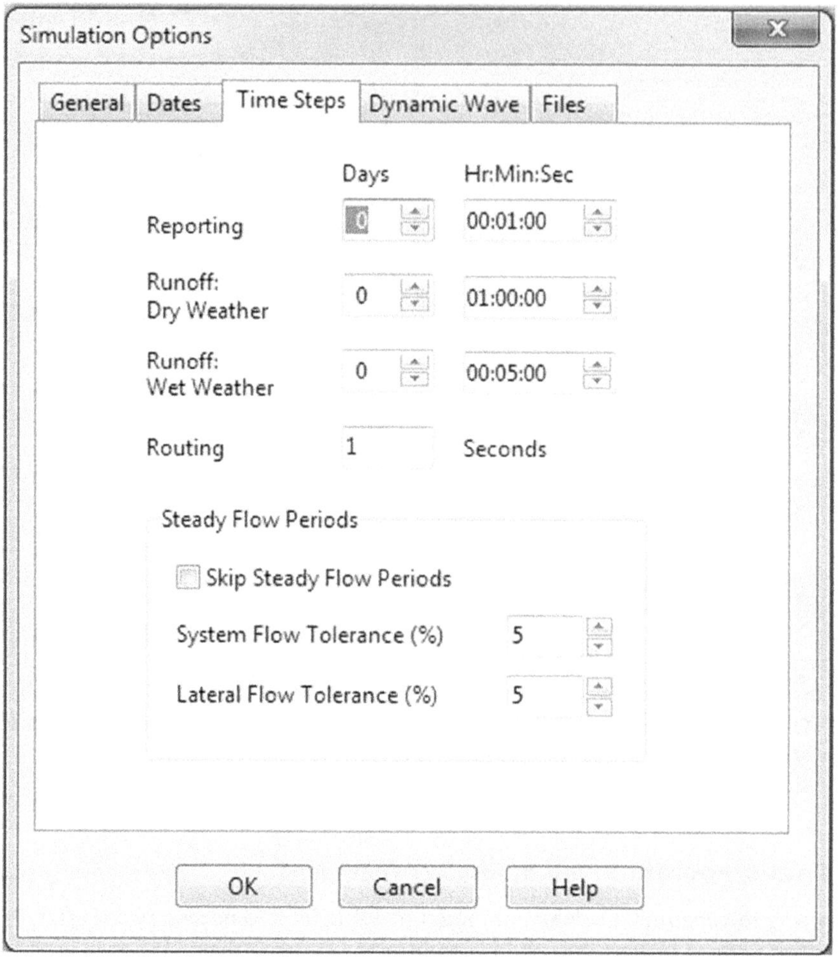

Figure 6.4 Dialog box for the step of saving calculations and solving differential equations.

characteristics of the catchment. In SWMM, the subcatchment (Figure 6.5) is described by the following parameters:

- catchment area (Area),
- flow path width (Width),
- mean longitudinal slope (Slope),
- percentage of impervious areas (%Impervious),
- percentage of impervious areas without initial retention,
- Manning's roughness coefficient for impervious (Nimp) and pervious (Nper) areas,
- depth of storage depression of impervious (Dstore-Imp) and pervious (Dstore-Perv) areas,
- number of the network connection node to which stormwater is discharged from the subcatchment (Outlet),

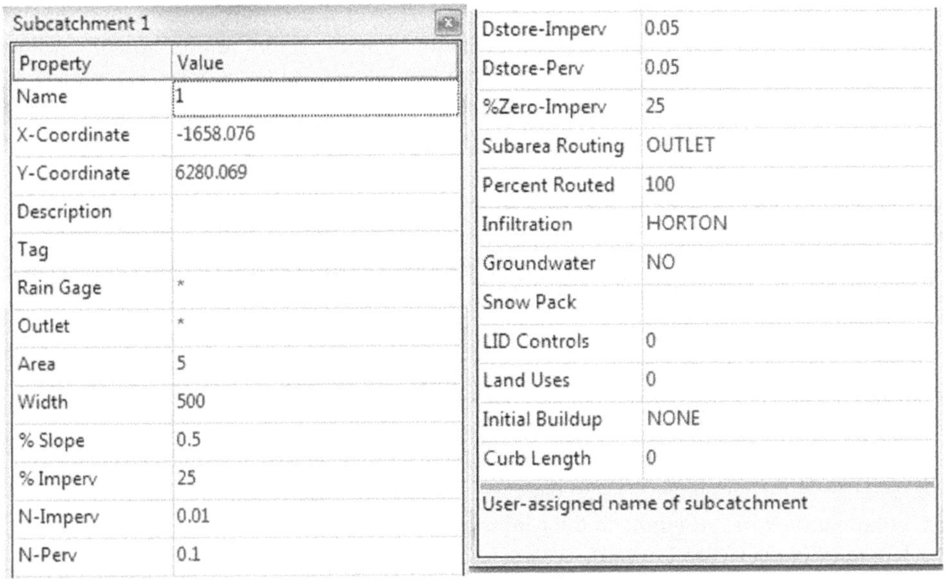

Figure 6.5 Parameters required for catchment modelling including analyses of stormwater depth and quality (upper and lower part of window).

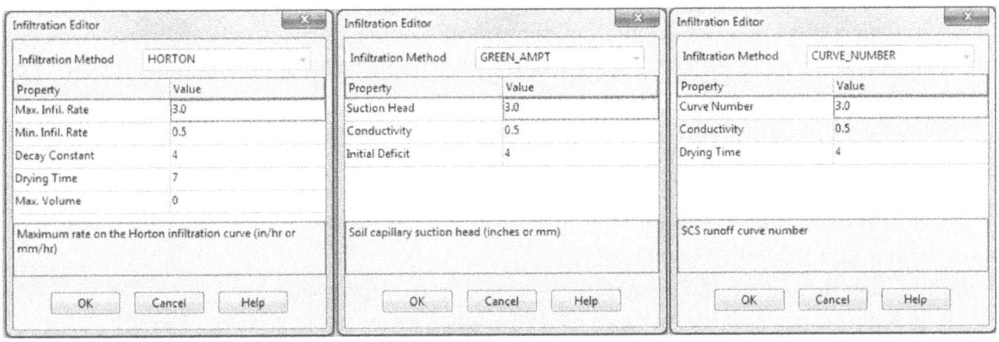

Figure 6.6 Dialog boxes for selected models of infiltration.

- parameters of the infiltration model; in SWMM, it is possible to simulate the process with the Horton or Green-Ampt methods or CN values (Figure 6.6),
- data on rainfall distribution in the analysed catchment (Rain gauge),
- scheme of rainfall water runoff in the catchment (Subarea).

There are several options in the SWMM program:

a runoff from impervious $Q_{imp}(t)$ and pervious $Q_{perv}(t)$ areas is independently directed to the sewer network ("Outlet" option), which allows calculation of the size of the stormwater stream $Q(t)=Q_{imp}(t)+Q_{perv}(t)$,

b stormwater flows from the impervious area $Q_{imp}(t)$ over a pervious surface, and the excess $F(Q_{imp}(t))$ flows into the sewer network ("Pervious" option); this results in lower values of flows in the channels, especially during intense rainfall events on an unsaturated ground.
c there is stormwater runoff from pervious areas $Q_{perv}(t)$ over impervious surfaces, and some amount flows into the sewer network $F(Q_{imp}(t))$ – "Impervious" option.

In the case of the subcatchment in SWMM, information on the proportion in the total catchment area and the solutions applied for rainwater management is entered in the LID model.

The SWMM program facilitates inclusion of data on the groundwater table (Groundwater) in calculations of catchment water balance; however, this requires field measurements. Thus, using this function in simulations without appropriate field studies is not advisable.

In addition to parameters (Figure 6.5) that have a significant impact on the dynamics of stormwater runoff from the catchment, it is possible to model stormwater quality. To this end, it is necessary to define:

- the land use in the catchment (Land use),
- the initial amount of pollutants in the catchment area (Initial Buildup),
- the length of curbs (Curb Length).

6.6.4 *Sewer network (manholes, channels)*

Manholes are an important element requiring determination of parameters during sewer network simulation (Figure 6.7a). Modelling of a runoff from an agricultural catchment with variable topography does not necessitate inclusion of "manhole" nodes, in which the inflows from catchments with higher location are summed up.

In the case of inclusion of manholes during the construction of the catchment model, it is necessary to define the following parameters:

- channel bottom ordinate (Invert El.),
- maximum filling depth (Max. Depth),
- initial filling depth (Initial Depth),
- filling depth resulting in stormwater overflow when exceeded (Surcharge Depth),
- area covered by overflow when the manhole maximum depth is exceeded (Ponded Area).

When the calculations are related to simulation of the operation of a combined sewer system in dry weather, the quantity of stormwater flowing into the wastewater treatment plant is highly variable. This fact is anticipated in the SWMM program; therefore, it is possible to introduce the inflow function for dry weather in nodes (manholes) (Figure 6.7b).

In these cases, it is necessary to specify:

- average value of flow intensity (Average Value),
- distribution of flow variability (monthly, daily, hourly, weekly).

Given the extensive possibilities of defining variability $Q(t)$ in nodes, it is feasible to create models of the sanitary sewer system in the SWMM program and thus assess the impact of

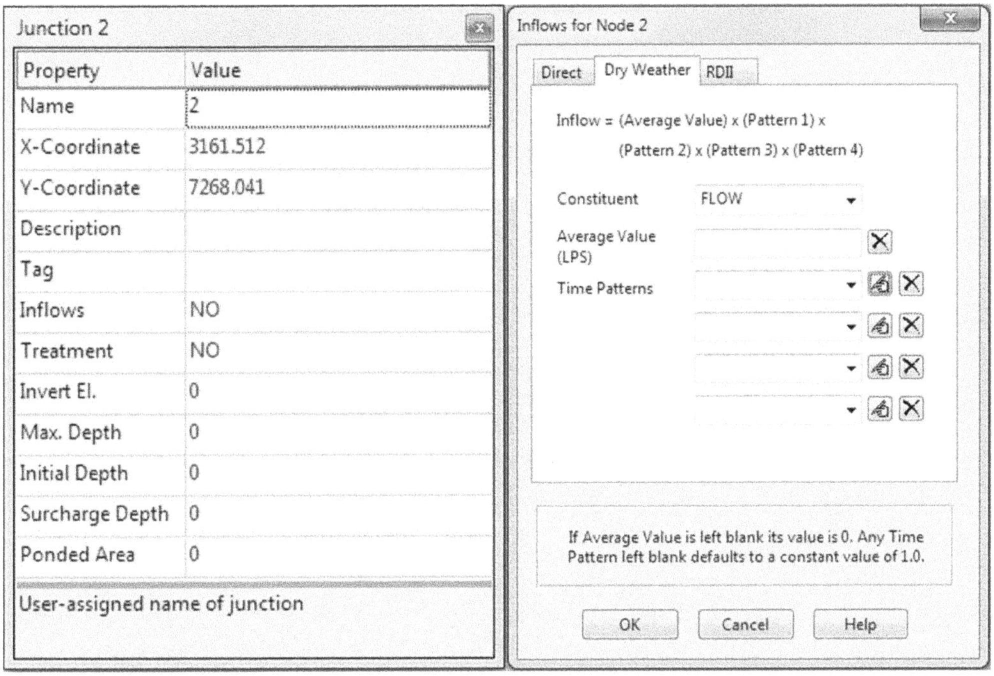

(a) (b)

Figure 6.7 Dialog box for (a) manhole parameters, (b) outflow from the node during dry weather.

intense rainfall on the operation of the sewer system and its facilities. This is possible by introducing the following function to the consecutive network nodes (connection manholes - inlet):

$$Q(t) = f(t) \cdot Q_0 \tag{6.29}$$

where: Q_0 – initial flow value (Baseline), $f(t)$ – values of coefficients of the distribution of stormwater inflow variability in the sewer network depending on the type of water use (Baseline Pattern, Figure 6.8).

In this way, it is possible to model the variability of the filling and flow velocity in sanitary sewers in settlement units taking into account the diversity of housing development (high, low, single-family), industrial plants, service development, etc.

Channels are another key element in the sewer network. The following parameters have to be defined to carry out simulations in the SWMM program (Figures 6.9 and 6.10):

- channel cross-section (Shape); the program offers a wide selection of channel cross-sections (parabolic, egg-shaped, bell-shaped, pear-shaped, trapezoidal, triangular, rectangular, channels with deposited sludge, irregular) - this gives a possibility of introducing a river bed,
- channel diameter (Diameter),
- channel length (Length),
- Manning's Roughness coefficient (Roughness),

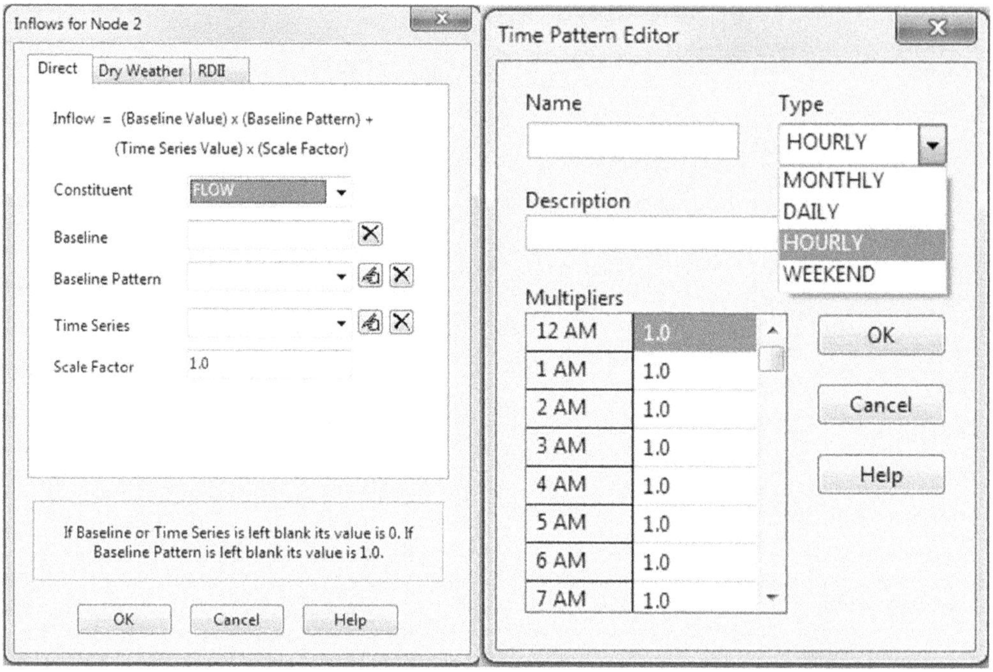

Figure 6.8 Dialog box for the variability of stormwater outflow from the node.

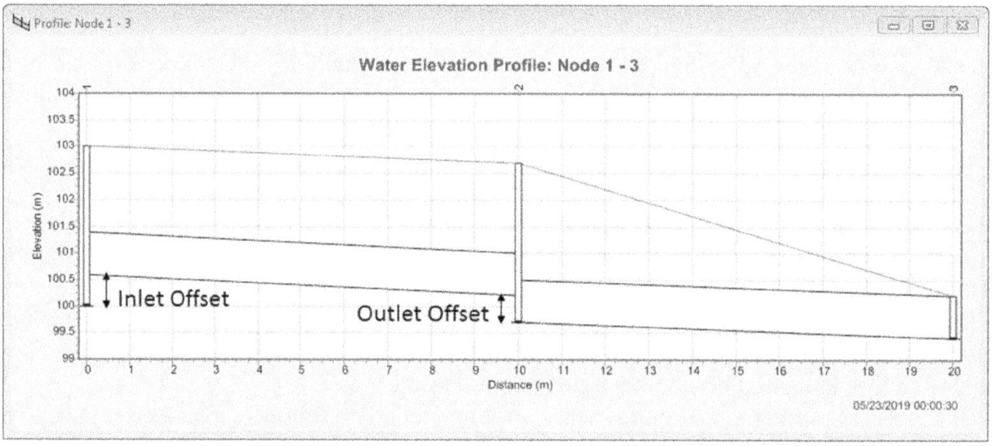

Figure 6.9 Visualisation of the "Inlet Offset" and "Outlet Offset" function.

- difference between the ordinates of the channel bottom at the inlet (Inlet Offset) and the outlet (Outlet Offset),
- initial flow values (Initial Flow),
- maximum flow values (Maximum Flow),
- local resistance coefficient at the channel inlet (Entry Loss Coeff),

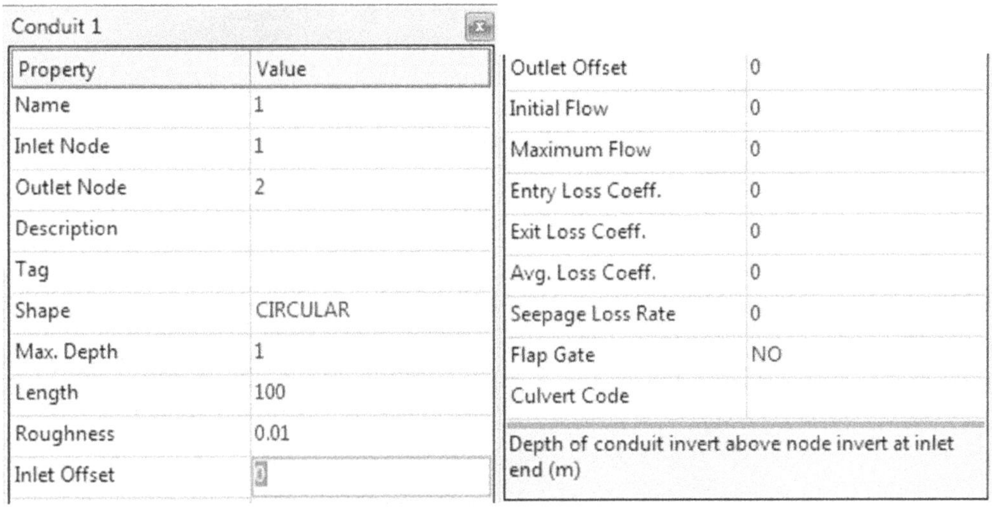

Figure 6.10 Parameters of a single channel section in the SWMM (upper and lower part of window).

- local resistance coefficient at the channel outlet (Exit Loss Coeff),
- volume of stormwater infiltration into the ground (Seepage Loss Rate).

6.6.5 *Storm overflow*

The SWMM program gives a possibility of simulating the operation of a storm overflow, i.e. the overflow depth and the instantaneous stormwater flow velocity. The program includes the following overflow design solutions:

$$\text{Transverse}: Q = C \cdot L \cdot (h - H)^{3/2} \tag{6.30}$$

$$\text{V} - \text{Notch}: Q = C \cdot S \cdot (h - H)^{5/2} \tag{6.31}$$

$$\text{Trapezoidal}: Q = C \cdot L \cdot (h - H)^{3/2} + C_{\text{WS}} \cdot S \cdot (h - H)^{5/2} \tag{6.32}$$

$$\text{SideFlow}: Q = C \cdot S \cdot (h - H)^{5/3} \tag{6.33}$$

In each of these cases, the following parameters have to be defined: Length (L), Height, edge slope (trapezoidal discharge, S), distance between the overflow edge and the bottom of the manhole (separation chamber, tank) (Inlet Offset H), discharge coefficient (Discharge Coeff., C) determined from Table 6.5, and an additional discharge coefficient for the side edges of the trapezoidal overflow (End Coeff.).

Used settings indicate that the kinetic energy of the stormwater stream entering the overflow is neglected at the stage of determination of the flow intensity. This implies that the Q values may be underestimated for the measured h–H value, which influences the reliability of value C determined at the model calibration stage.

Table 6.5 Values of overflow discharge coefficients (Rossman 2004; Rossman 2009)

Type of overflow	*C*
Frontal with a sharp edge	1.84
Frontal with a wide crown	1.38 ÷ 1.83
Frontal triangular	1.35 ÷ 1.55

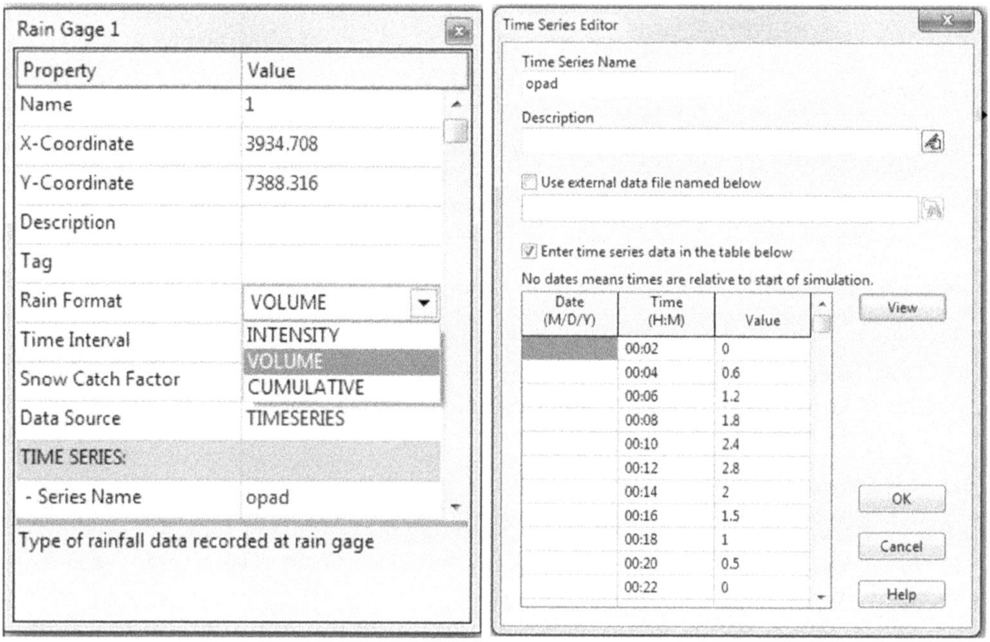

Figure 6.11 Dialog boxes for rainfall data

6.6.6 *Rainfall*

Rainfall is a key variable in the simulation of the operation of a stormwater and combined sewer network. In the SWMM, rainfall can be defined as continuous time series or single rainfall events. The basic variables describing rainfall (which must necessarily be defined) include (Figure 6.11):

- time step (Time Interval),
- rainfall depth, which can be expressed as rainfall intensity (Intensity), rainfall volume (Volume), and cumulative rainfall (Cumulative),
- unit (Unit) expressing the rainfall.

The case discussed above describes a situation when rainfall covers the catchment area. However, when the front moves over the catchment area at a speed v, not all catchments are covered

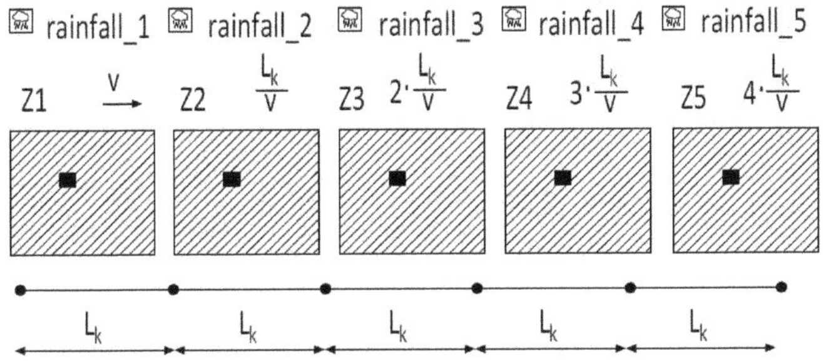

Figure 6.12 Calculation scheme in a model including movement of the weather front.

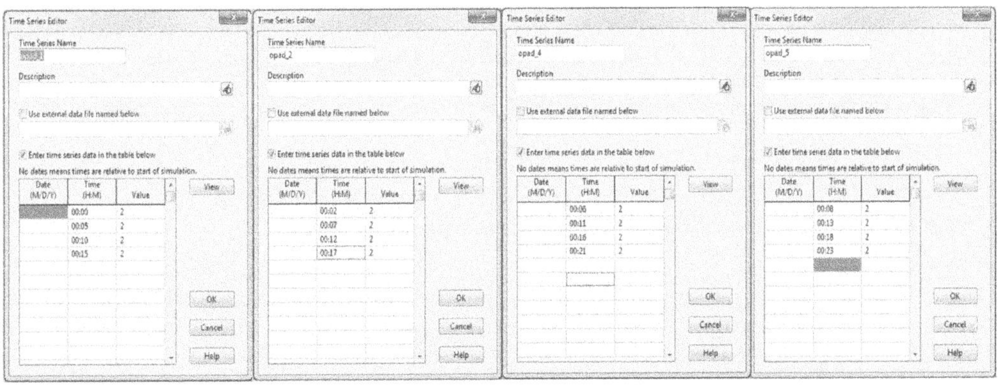

Figure 6.13 Distributions of rainfall in successive subcatchments of the model.

by rainfall. For such a case, a calculation scheme was generated (Figure 6.12) to include the rainfall data in the constructed model correctly. Knowing that the length of the catchment area is similar to the channel length L_k, it was established that the time of front movement over a subcatchment is $L_k \cdot v^{-1}$. This indicates that the rainfall will reach catchment Z3 after time $2 \cdot L_k \cdot v^{-1}$ and consecutive catchments Z4 and Z5 after time $3 \cdot L_k \cdot v^{-1}$ and $4 \cdot L_k \cdot v^{-1}$.

The resulting rainfall hyetographs for the assumed rainfall distribution and $L_k \cdot v^{-1} = 2$ minute are shown in Figure 6.13.

6.6.7 *Integration of the canal network model with road surfaces*

At the stage of modelling sewer systems, it is essential to analyse the conditions of operation of channels in the event of stormwater flooding onto the ground surface. Thorough understanding of such complex phenomena is a key issue for making appropriate decisions on modernisation of the sewer system. The SWMM program facilitates stimulation of the phenomenon of stormwater flooding only in the area of roads and parking lots. To capture this phenomenon at the modelling stage, a double model has to be generated to include sewer pipes and road cross-sections between successive manholes (Figures 6.14 and 6.15).

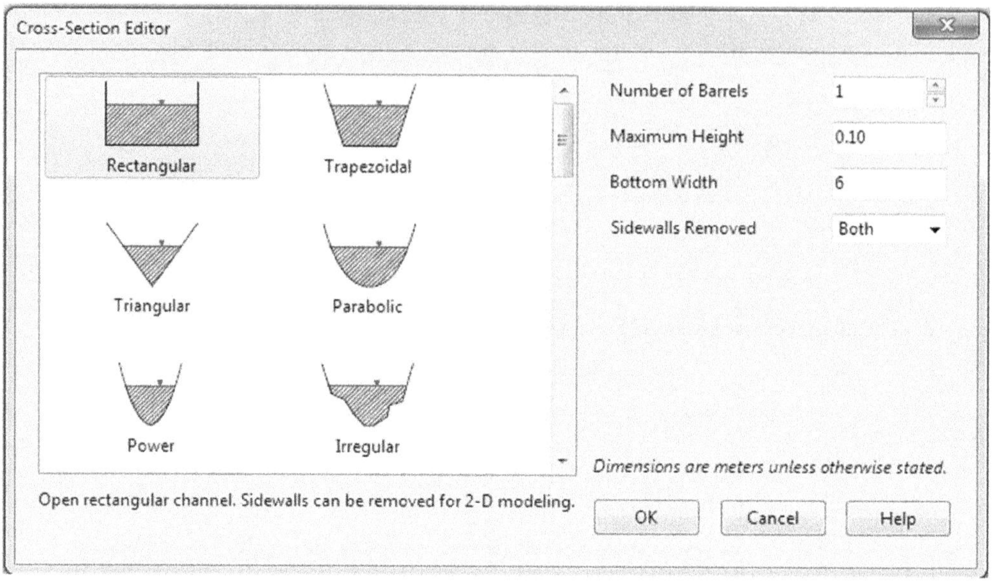

Figure 6.14 Cross-section of a road between overflowed manholes.

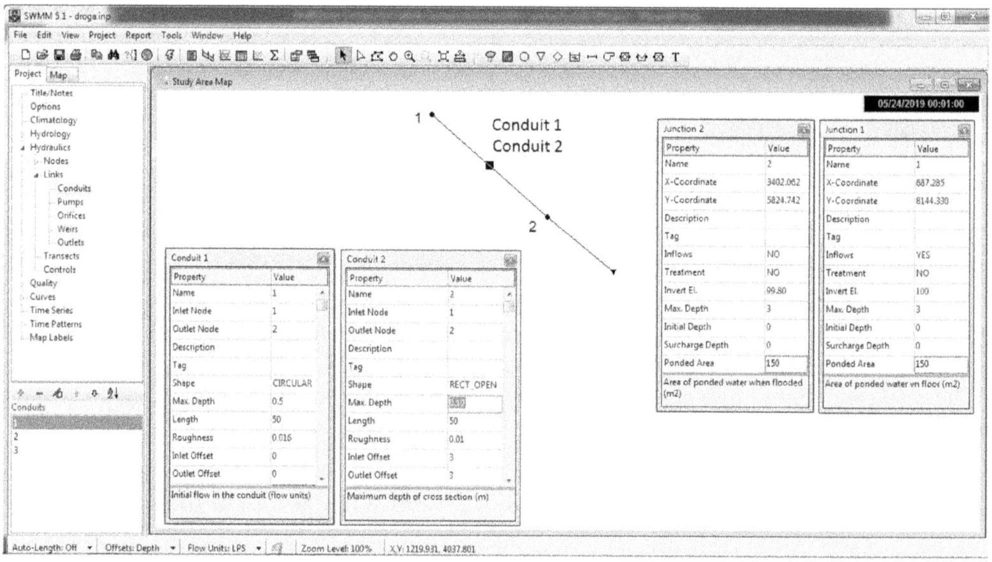

Figure 6.15 Calculation diagram for a sewer system and the road surface with stormwater flood.

6.6.8 *Wastewater pumping stations*

Pumping stations, which in many cases are an integral element of retention tanks, are important objects in sewer networks (stormwater, sanitary). The SWMM program assumes various types of variability $H=f(Q)$; however, under technical conditions, the simulation of the pump station is based on the characteristics of the pump, which can be entered in the Pump curves>Pump curves editor tab (Figure 6.16).

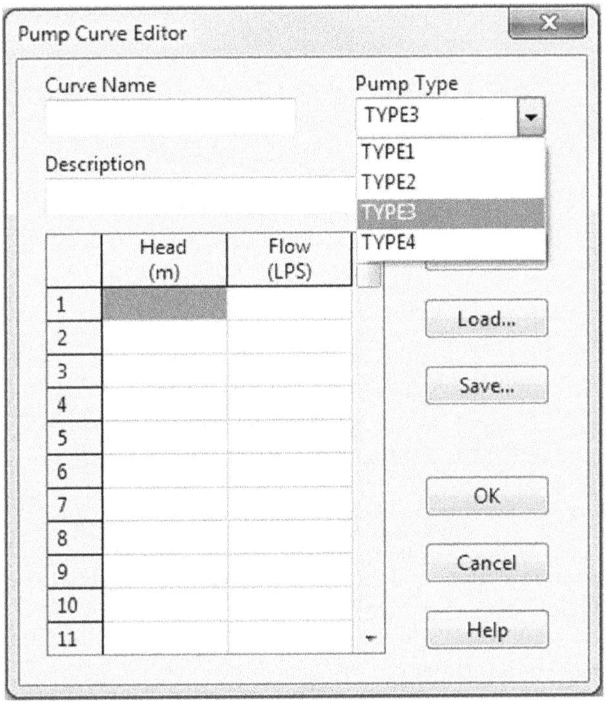

Figure 6.16 Dialog boxes for pump characteristics.

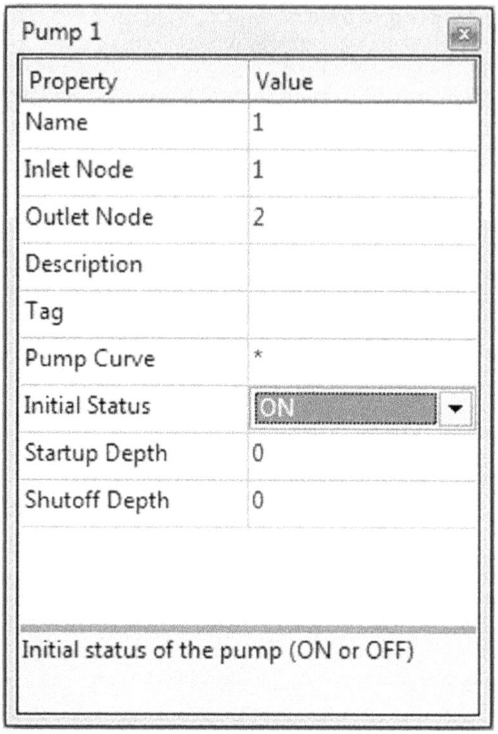

Figure 6.17 Dialog box for data on pump characteristics.

Generation of a pumping station model in the SWMM program should be preceded by definition of the following parameters (Figure 6.17):

- pump start-up level (Startup Depth),
- pump shut-off level (Shutoff Depth).

On the basis of the data presented above, the SWMM program generates models of sewer networks with pumping stations.

Chapter 7

Runoff regulation and control

7.1 Possibilities of reducing wastewater impact of on urbanised areas and receivers

This chapter discusses the possibilities of reducing the negative impact of discharged wastewater, both on urban agglomerations and on receivers. It is necessary that all measures, whether implemented in the urban catchment or in the sewerage network, pursue a single objective, namely to reduce the amount of pollution discharged to recipients from CSOs and wastewater treatment plants.

In particular, the following shall be considered as measures to reduce the amount of wastewater discharged from CSOs and surface runoff, as well as to reduce the amount of pollution discharged by such water:

- separate drainage of surface runoff and wastewater,
- use of surface retention of the urban catchment and its self-cleaning ability,
- infiltration of low-polluted water from surface runoff into the ground,
- semi-vegetation and vegetation measures,
- utilisation of the retention capacity of the sewer network,
- inclusion of rain tanks or vortex separators in the sewer system,
- construction of progressive CSO types with pollution reduction (bars, screens, floating matter and debris traps/barriers, down flow baffle systems etc.),
- flows real time control (RTC) in the sewer network,
- inclusion of sediment traps and traps for non-polar extractable substances (oils, petrol),
- cleaning of rain gutter collection devices,
- flushing and cleaning of sewer sections with excessive sediment formation,
- a combination of previous measures.

Captured pollutants must be taken to a wastewater treatment plant or disposed of in another suitable way. Depending on the location of the procedure, two options for regulating the outflow are distinguished.

7.1.1 Regulation of runoff formation

It concerns regulation, or reduction of the amount of outflow, delay, or reduction of peak flow, redirection of runoff to the catchment of another channel, construction of open stormwater tanks, etc. The largest group of regulatory measures of this group consists of infiltration devices

DOI: 10.1201/9781003546436-7

of various types (porous surfaces, gutters, trenches, manholes with pervious bottoms, infiltration pipes, pervious pavements and curbs, rain gardens). They can be used especially in locations with individual buildings.

7.1.2 *Regulation of outflow in the stormwater network*

This group of regulatory measures includes the construction of underground stormwater tanks, the construction of regulatory elements in the stormwater network (e.g., overflows, sluice gates, closures), rain separators, modified inlets with retention space, etc. Regulation of runoff in the stormwater network often requires verification of the operation of the designed devices by hydrodynamic modelling of stormwater runoff.

7.2 Decentral measures (Runoff generation regulation)

7.2.1 *Runoff reduction*

In terms of infiltration, green roofs and rain gardens below are listed the most used regulatory measures:

- infiltration,
- retention on flat roofs,
- retention on green roofs,
- retention in downspouts and subsequently in barrels under the downspouts,
- drains with a retention space or installed at a greater distance from each other,
- surface flooding (especially parks and parking lots).

7.2.2 *Infiltration*

Infiltration measures can be divided into two basic groups:
Surface infiltration

a surface infiltration: grassy surfaces, semi-vegetated concrete prefabs or other porous surfaces e.g., water permeable concretes,
b concentrated infiltration: open stormwater tanks and infiltration gutters,

Subsurface infiltration

a concentrated infiltration: infiltration wells (dry in the dry season) their construction is similar to sewer shafts with a different design of the bottom and sometimes also the side walls,
b line infiltration: infiltration trenches with a perforated pipe distribution system within the infiltration trench filled with gravel or quarry stone. Individual methods can be used alone or in combination with others, e.g., infiltration wells together with infiltration gutters.

7.2.2.1 *Surface infiltration*

Surface infiltration stormwater disposal requires porous surfaces and soil-filled precast concrete to have a hydraulic conductivity greater than or equal to the design rainfall intensity.

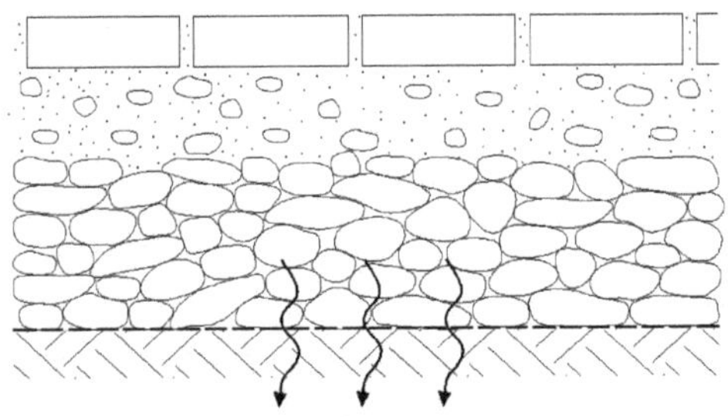

Figure 7.1 Cross-section of a pervious parking lot.

These surfaces are most often designed for parking lots, local roads, or storage areas (Figure 7.1). Such a solution enables the immediate absorption of stormwater into the subsoil, and, in addition, pervious parking lots can also receive water from the roofs. It is advantageous if these surfaces are designed without curbs, or using interrupted curbs, so that, in case of heavy torrential rains, when the subsoil is saturated with water, the stormwater can flow freely into the gutter, or to the adjacent grassy area. Over time, preferential paths of filtering water through the filter layer are often created, which then become easily clogged. Therefore, it is necessary to pay maximum attention when choosing materials to their greatest possible homogeneity, so that zones with different densities are not created during construction. The essence of the technical implementation of surface infiltration consists in the fact that a supporting layer of gravel or sand is spread on the pervious subsoil, on which either concrete prefabs are laid (the holes of which can be filled with gravel or soil and can be grassed) or a road surface is created with porous asphalt covering on a macadam base. The use of these pervious surfaces, as well as other infiltration devices, is not universal and depends mainly on local hydrogeological conditions. Pervious areas are particularly suitable for drainage of smaller parking areas, storage areas, and driveways.

Clogging of the surface layers with runoff requires their replacement after several years of operation. The expected lifespan is estimated at 10–15 years. When using grass prefabs, renewed infiltration can be realised by lifting the concrete tiles, removing the gravel layer and geotextile, and replacing them with a new layer of gravel and new geotextile.

The advantage of these areas is that, during the winter months, they do not need to be de-iced. When snow melts, the water drains directly into the underlying layers, which prevents the formation of ice.

7.2.2.2 *Surface infiltration gutters*

Purpose-designed surface infiltration gutters are commonly used in North America for surface stormwater retention. They are grassy trenches with small slopes of 1:3, which facilitate mowing. Stormwater is brought to the gutter using stormwater gutters or directly drains to the side of the slope. The depth of the manholes usually does not exceed 1 metre. Catchment trenches are designed up to 5% longitudinal slope. For individual construction, these gutters can also be made as small depressions on the grass area (Figure 7.2). A great advantage of this method of infiltration is simple construction and especially easy maintenance.

In addition to regulating the amount of runoff, gutters also reduce the amount of pollution, because sedimentation and chemical adsorption take place in them. In this way, they reduce the

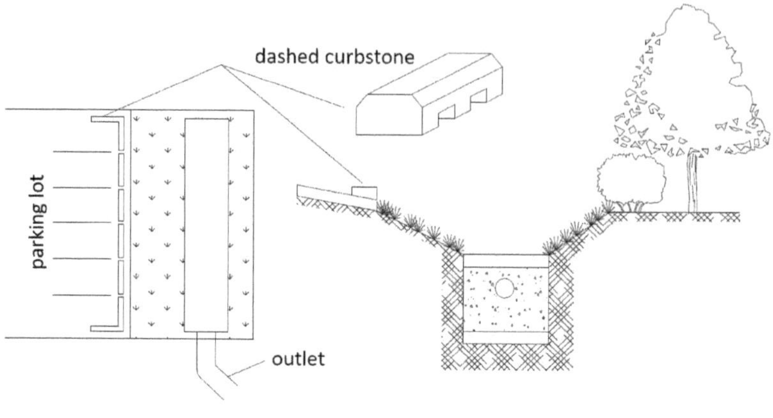

Figure 7.2 Scheme of the infiltration gutters.

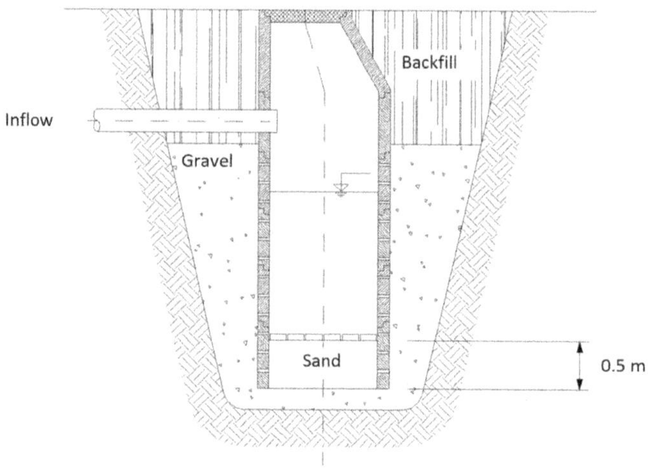

Figure 7.3 Infiltration well.

amount of insoluble substances, and, thus, also the amount of other pollutants adsorbed on the sediments (e.g., heavy metals), which could contaminate the infiltration wells or, ultimately, the groundwater.

7.2.2.3 *Infiltration wells*

Infiltration wells are compact infiltration devices that can be built from bricks, concrete, concrete rings, or from quarry stone on private land, or in connection with infiltration trenches, also for the infiltration of larger amounts of stormwater, e.g., in business premises (Figure 7.3). They can also be included as an absorption element before the stormwater enters the stormwater network.

During their design, care must be taken to select suitable material, to eliminate floating dirt using rakes or sieves, and care must be taken to avoid unauthorised use during operation (e.g., for soaking wastewater), to allow access for regular cleaning, to prevent their clogging by garbage or soil.

Infiltration wells are usually built to a depth of 2–3 m, but deeper ones are also known. These are examples where existing wells extending to greater depths were used for infiltration. When designing deep infiltration wells, the possibility of groundwater contamination must be taken into account. In Slovakia, such deep wells built by piercing up to the HPV level were used in the disposal of water during the design of the D61 highway.

7.2.2.4 *Infiltration trenches*

The principle of infiltration trenches is similar to the principle of infiltration wells. The inflow of stormwater can be directly through the surface from adjacent areas, or through a filter (Figure 7.4), or they can be fed through perforated pipes (Figure 7.5). In some countries, grassed areas are built in front of infiltration trenches for the sedimentation of insoluble substances found in stormwater.

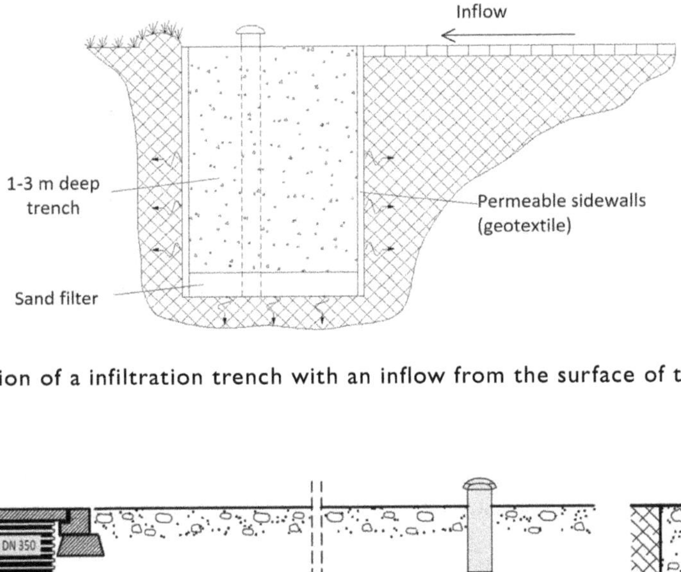

Figure 7.4 Section of a infiltration trench with an inflow from the surface of the parking lot.

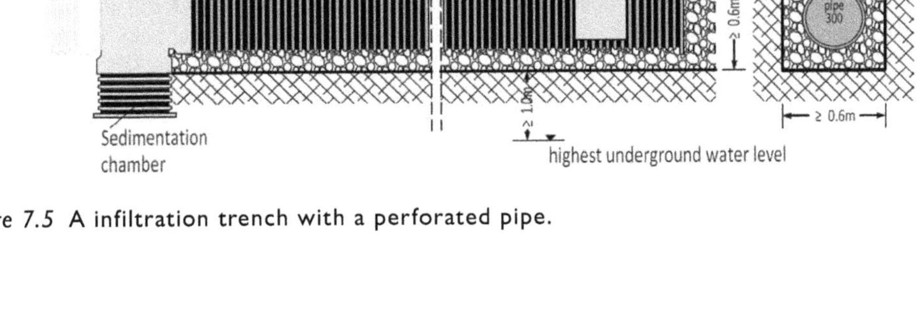

Figure 7.5 A infiltration trench with a perforated pipe.

The advantage of infiltration trenches over infiltration wells is that they provide a larger area for draining of trapped water below the surface. Narrower and longer infiltration trenches are more efficient than wells or circular and square trenches. Appropriate arrangement of these trenches can create an interconnected drainage system that creates greater security against property flooding. If necessary, the unabsorbed amount of stormwater runoff from the infiltration trenches can be released into the stormwater network using underground pipelines, thus ensuring a delay in the peak of the stormwater runoff wave.

Perforated pipes can also be laid in infiltration trenches, and longer trenches can be connected in shafts. In such cases, ventilation pipes should be placed at least every 50 m. Shafts can also be used to change the direction of the trench.

7.3 Central measures (Runoff regulation in sewer system)

7.3.1 Stormwater retention tanks

7.3.1.1 Stormwater retention tanks and their role in the sewer system

The construction and expansion of infrastructure in urban agglomerations and rural areas, the expanding urban areas, and the new transportation routes, i.e., highways and numerous asphalt roads, contribute to an increase in surface runoff. It is associated with the reduced area of water infiltration into the ground and lower surface and ground retention in the catchment. This increases the quantity of water flowing into stormwater sewer systems and receivers. In each solution targeted at drainage of roads and other facilities, the possibilities of stormwater management using natural objects for rainwater infiltration (basins, trenches, ponds, etc.) are analysed (Geiger and Dreistl 1999). Particular importance in the drainage system is assigned to the regulation of stormwater flow in the sewer system, which has a major impact on the dimensioning and rational use of the sewer network and protection of surface waters. Tanks intercept peak flows and temporarily store substantial amounts of stormwater, which is discharged into the receiver in a controlled manner. They also facilitate:

- connection of new catchments to the existing sewer network; in the case of new housing estates that have to be connected to the existing system, the retention capacity allows integration of stormwater without the necessity to reconstruct the existing sewer network,
- an increase in the hydraulic capacity reserve of the existing sewer network with concomitant reduction of the probability of network overfilling during torrential rainfall,
- retention of excess stormwater discharged through stormwater sewers to receiver waters and reduction of the intensity of stormwater outflow from the stormwater sewer system,
- hydraulic unloading of the stormwater sewer system and its facilities and devices located below the reservoir – it is now possible to use an optimal design of a tank ensuring controlled outflow, even when the capacity of the channels located below or the capacity of the receiver are insufficient,
- reduction of the discharge, number of pumps, and volume of stormwater or industrial sewage pumping stations located in the sewer network below the tank,
- protection of receiver waters and unloading technological facilities cooperating with stormwater and municipal wastewater treatment plants in a system with storm overflows (Dąbrowski 2004),
- averaging the composition of sewage flowing to the municipal wastewater treatment plant in combined and stormwater sewer systems (Figure 7.6),

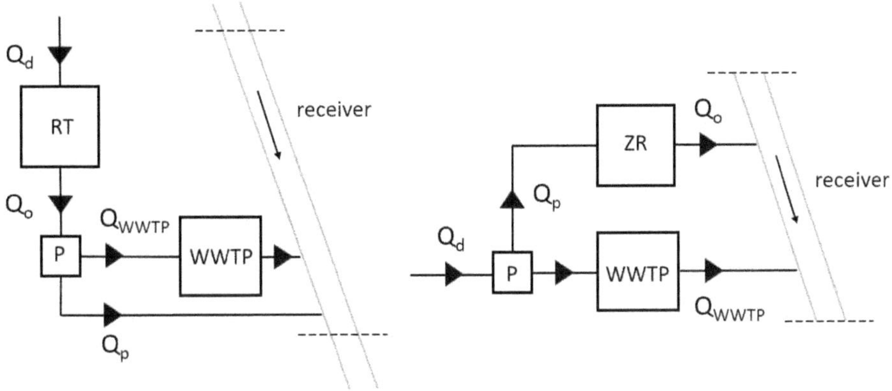

Figure 7.6 Examples of solutions for sewer systems with a retention tank.

Notes: RT – retention tank, WWTP – wastewater treatment plant, *P* – outflow, Q_d – intensity of inflowing stormwater, Q_o – intensity of stormwater outflow from the retention tank, Q_p – intensity of stormwater outflow from the storm overflow, Q_{WWTP} – intensity of stormwater outflow from the wastewater treatment plant.

- design of a smaller diameter of channels located below the tank outlet – it is possible to correlate the cross-section of channels discharging stormwater from the tank with the capacity of the existing sewer system, the depth of the channel foundation, and the groundwater level. Moreover, it is possible to renovate channels in the existing sewer system in order to increase their capacity without the need to enlarge the channels,
- protection of the municipal sewer system during periodic discharges of various types of industrial wastewater (Dziopak 2004),
- elimination of the necessity to destroy the existing surface of roads and squares during reconstruction of the stormwater sewer network (enlargement of the channel diameter) under urban conditions. This results from the fact that excess stormwater flowing into the channels is accumulated in retention tanks and the sewer network is hydraulically unloaded,
- acceleration of the process of residential construction investment owing to the substantial reduction in the costs and time of construction of underground stormwater networks.

There are two basic solutions of sewer systems cooperating with storm overflows and the tank. In the first solution, the tank is located before the storm overflow (Figure 7.6, left side), whereas the second solution assumes location of the storm overflow before the tank (Figure 7.6, right side). In these solutions, it is possible to use appropriate tank structures that facilitate reduction of the flow rate and pre-treatment of wastewater flowing into the WWTP (Słyś 2009) in the process of sedimentation, flotation, and thickening of sludge (Tabernacki 1980). Moreover, the installation of the tank before the storm overflow reduces the frequency of discharges from the latter and limits its dimensions by shortening or lowering the overflow edge relative to the inlet channel bottom. Other variants of the designs and their functions are discussed extensively in the textbooks by Dziopak and Słyś (2006; 2007).

Retention tanks are used in the country (Germany, Austria, Italy, Poland, etc) although they contribute to reduction of the size and cost of channels and other sewer network facilities located below the tank, improve the effects of wastewater treatment plants, and facilitate protection of receiving waters. This is primarily associated with the construction costs, which are influenced by the receiver water quality and level, the groundwater level, the capacity of channels and/or

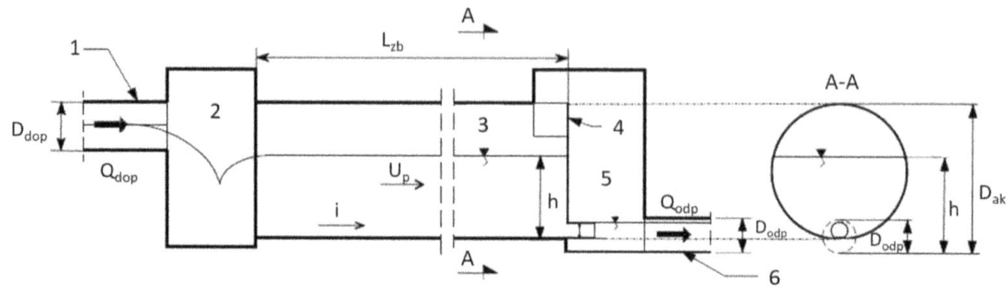

Figure 7.7 Scheme of a tubular tank. (Based on Szeląg et al. 2021b).

Notes: D_{ak} – diameter of the retention chamber, L_{zb} – length of the retention chamber, D_{sp} – diameter of the tubular chamber orifice, D_{dop} – diameter of the inlet channel, D_{odp} – diameter of the outlet channel, h – accumulation chamber filling, i – longitudinal slope of the tubular chamber, U_p – average velocity of stormwater flow in the retention chamber, Q_{dop} – stormwater stream flowing into the tank, Q_{odp} – stormwater stream flowing out of the accumulation chamber; 1 – inflow channel, 2 – inlet chamber, 3 – retention chamber, 4 – emergency overflow, 5 – outlet chamber.

wastewater treatment plants below the tank, and the geology, topography, and degree of urbanisation of the area.

The least complicated group of underground retention tanks in stormwater and combined sewer systems are structures with one accumulation chamber (single-chamber), in which the following relief devices are used: a pipeline equipped with a gate valve for regulation of the flow (drain pipes, reducers, flow regulators, pumps, etc.). The type of the relief device applied depends on the function served by the tank and the required maximum flow reduction effect (Q_{dmax}). In the classic solution (cuboidal accumulation chamber and outflow via the drain hole), the designed tank requires a large projection area, which often limits the possibilities of construction thereof in the catchment and necessitates a new appropriate construction design. A tubular tank (Figure 7.7) has no such drawbacks; nevertheless, similarly to the classic tank, it is characterised by low efficiency and effectiveness of the accumulation chamber and hydraulic capacity at the stage of wastewater retention. In this solution, the inflow and outflow to and from the tank proceed through inlet and outlet chambers, respectively. The advantages of single-chamber tanks prompted the search for other solutions aimed at reduction of the space required for the construction of the tank, reduction of construction costs, and simplicity of operation with the tank self-cleaning option (Dziopak 1986; 1988; 1997). Investigations resulted in the development of innovative multi-chamber gravity tanks differing in the location of accumulation chambers (Dziopak 2000). The solution was based on the use of a flow chamber serving as a flow regulator, facilitating maintenance of a constant outflow rate during the tank filling stage and a substantial reduction in the tank capacity – in extreme cases even by 70% – in comparison with the single-chamber tank (Dziopak 1997). Additionally, contaminants dragged into the wastewater stream in the inflow channel do not reach the retention chambers but flow through the flow chamber into the sewer system below the tank and into the wastewater treatment plant.

Despite the significant reduction of their volume, multi-chamber gravity tanks still required a large area for construction thereof, and the sewer network had to be located at a larger depth below the tank due to the location of the outflow channel. Further search for solutions of construction of tanks was targeted at reduction of their surface area and maximum shallowness of the outflow channel. One of the first solutions meeting these criteria was the design of gravity-vacuum tanks (Kisiel 1998; 2005a,b). The use of kinetic energy of stormwater flowing out through the drainage channel, and/or assisted by vacuum pumps could facilitate accumulation of wastewater at a level higher than the inflow channel ordinate, extension of outflow time and intensity, and reduction of the depth of outflow channel foundation. Unfortunately, the necessity to build a covered structure ensuring tightness of all elements affected by negative pressure, and the low efficiency of jet pumps substantially increase the investment and operating costs (Kisiel 1998). These factors contributed to initiation of research on cheaper solutions for accumulation of wastewater at a level higher than the ordinate of the inflow channel.

The use of pumping systems (Dziopak & Słyś 2007) for wastewater transfer in the multi-chamber tank contributed to reduction of its surface, location of the inlet and outlet at a similar level, and limitation of expenditure on construction of chambers (its operation does not require covered structures). Horizontal multi-chamber tubular tanks have been developed as well (Kisiel 1998, Kisiel et al. 2008). The hydraulic capacity of the proposed solutions was comparable to that of classic multi-chamber structures. The essence of the developed patent solutions for tubular tanks consisted in the use of an additional partition separating the accumulation chamber (KA) from the inlet chamber (KWl), which in the bottom zone has an inlet to the transit channel running along the accumulation chamber (KA). The stormwater outflows from KA through the orifice (OS) located at the lowest point of the tank chamber. The stormwater flow rate through the drain hole is regulated by a float valve (ZP), the discharge of which depends on the filling of the outlet chamber (KWy). In another design of a multi-chamber tubular tank, the float regulator was eliminated, which increased the operational reliability of the tank.

Given the retention tank structures described above, it can be concluded that the use of a flow chamber serving as a flow regulator was highly important for the development of these tanks. This is confirmed by the fact that gravity-pump, gravity-vacuum, transfer, and gravity tubular tanks have a flow chamber in their design. These tanks represent a more advanced solution than single-chamber tanks due to the shallower location of the outflow channel (allowing stormwater discharge to shallow receivers) and the reduced accumulation capacity of the tank.

In the case of single-chamber tanks, tubular structures deserve special attention. Although they have the same shortcomings as classic tanks, they can more frequently be installed in urbanised areas, as they are easy to construct and occupy small areas; hence, they can be used as an alternative system of multiple tanks located within the sewer network instead of one collective system.

7.3.1.2 *Theoretical basis for design of the capacity of retention tanks*

The main parameter of the operation of the stormwater retention tank is the maximum flow reduction factor (β) described by the following equation (Błaszczyk et al. 1974):

$$\beta = \frac{Q_{o\max}}{Q_{d\max}} \tag{7.1}$$

where: Q_{omax} – maximum outflow from the retention tank (m^3/s), Q_{dmax} – maximum inflow to the retention tank (m^3/s).

When the main task of the tank is to reduce the costs of construction and operation of sewer systems, β is determined based on the economic analysis of the costs of construction and

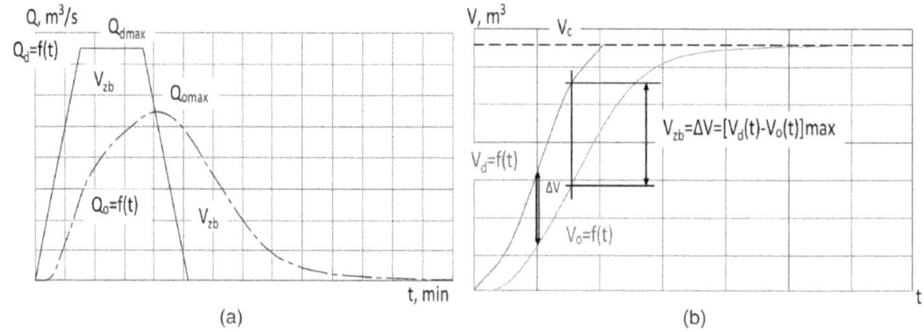

Figure 7.8 Methods for calculation of the capacity of the retention tanks based on inflow and outflow hydrograph (left), inflow and outflow mass curves (right).

Notes: $Q_d = f(t)$ – inflow hydrograph, $Q_o = f(t)$ – outflow hydrograph, Q_{dmax} – maximum wastewater inflow rate, Q_{omax} – maximum wastewater outflow rate, $V_d = f(t)$ – mass inflow curve, $V_o = f(t)$ – mass outflow curve, V_{zb} – tank capacity, ΔV – instantaneous difference in the ordinates of mass inflow and outflow curves.

operation of the retention tank with the sewer network and its facilities located below the tank, assuming different values of the flow reduction coefficient. In turn, when the tank is meant to protect the sewer facilities from overload, the determination of β is based on the maximum outflow from the tank protecting the sewer network or its part located below the tank. If the main purpose of the tank is to protect the receiver waters from excessive contamination with stormwater or mixed sewage, β is determined with the value of the maximum allowable stormwater or combined sewage inflow (with a known concentration of pollutants) to the receiver.

Methods for dimensioning retention tanks are based on the hydrological principle of comparison of the course of inflows and outflows over time. One of the methods consists in determination of the surface area of the Figure 7.8 outlined by the hydrograph of retention tank inflows and outflows (Figure 7.8, left). Another method is a modification of the former one and involves preparation of mass inflow and outflow curves (Figure 7.8, right).

The theoretical basis for calculation of tank capacity is the Saint Venant system of differential equations (Eqs. 7.30 and 7.31) for mass and momentum conservation. These equations are used for detailed analysis of the variability of the stormwater stream and the stormwater depth (h) at time t along the tank length (l). The full system of equations is the basis for assessment of the functioning of dam retention reservoirs, whose operation is largely influenced by the cross-section along their length. The tanks in the stormwater sewer system have much smaller sizes, and their stormwater table can be ignored, which gives the following relationship:

$$\frac{\partial Q}{\partial l} + \frac{\partial A}{\partial t} = 0 \tag{7.2}$$

which is transformed to the following equation after integration within limits l_1 to l_2:

$$Q(l_1) - Q(l_2) + \frac{d}{dt}\int_{l_1}^{l_2} A\,dl = 0 \tag{7.3}$$

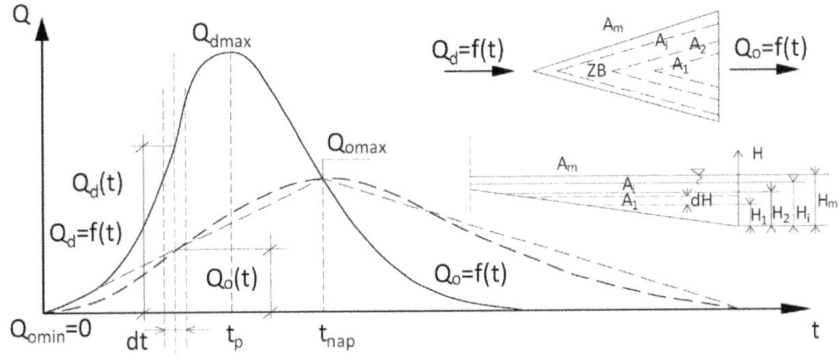

Figure 7.9 Example of an inflow and outflow hydrograph for a single-chamber retention tank.

where: l_1 and l_2 – coordinates at the inlet and outlet of the tank, respectively, $Q(l_1)$ – inflow into the tank, $Q(l_2)$ – outflow from the tank.

Since the value of integral $\int A dl$ describes the volume of stormwater accumulated in the tank, Eq. 7.3 can be written as follows

$$\left(Q_d(t) - Q_o(t)\right) = \frac{dV}{dt} = A(h) \cdot \frac{dh}{dt} \tag{7.4}$$

where: $Q_d(t)$ – rate of inflow into the tank (m^3/s), $Q_o(t)$ – rate of outflow from the tank (m^3/s), dh – increase in tank filling over time dt, dV – increase in wastewater volume in the tank over time dt, $A(h)$ – tank surface area (m^2) generally depending on the filling. The variables included in Eq. 7.4 are illustrated in Figure 7.9.

Due to the complex form of the inflow hydrograph, the solution of the differential volume balance equation is not simple, as it requires appropriate numerical algorithms (Szymkiewicz 2000). Nevertheless, simplified methods have been developed, in which a triangular scheme of the outflow hydrograph is assumed or Eq. 7.4 is transformed into finite differences. Additionally, to avoid solving the equation (Eq. 7.2), a number of methods have been proposed based on the numerical solution of the differential volume balance equation, in which a nomogram based on the results obtained serves for determination of the required accumulation capacity of the tank (Szeląg & Mrowiec 2016).

7.3.1.3 *Methods for calculation of retention tank capacity*

The differential equation of stormwater volume balance is the basis for assessment of the functioning of the retention tank. However, due to the complexity of the surface runoff phenomenon, the description of the dynamics of stormwater inflow to the tank $Q_d(t)$ is complicated, which makes the solution of Eq. 7.4 difficult. Therefore, numerous simplifications have been developed for the description of the inflow hydrograph shape or the output of relief devices, which facilitate the solution of the differential volume balance equation. Hence, the tank dimensioning approaches can be divided into the following methods:

- simplified, in which the outflow from the tank is described by a linear relationship yielding a triangular or trapezoidal outflow hydrograph,
- graphical-analytical based on a transformed differential equation of stormwater volume balance,
- graphical based on the numerical solution of Eq. 7.4 followed by derivation of combined parameters of the inflow hydrograph and hydraulic characteristics of the tank,
- numerical based on algorithms for solving Eq. 7.4,
- stochastic based on distributions of probability density developed based on multiyear rainfall measurement results or catchment outflow hydrographs.

7.3.1.4 Simplified methods

In the simplified methods, the calculation of the capacity of the retention tank can be divided into two stages. The inflow hydrograph is determined in the first stage, while the second stage involves determination of the value of the unit capacity of the tank (Annen & Londong 1966, Pecher 1972, Błaszczyk et al. 1974, Dziopak 1984, Mrowiec 2009). The tank capacity can be determined with the simplified methods based on:

- unit capacity index (W) showing the capacity of the designed retention tank per 1 ha of reduced catchment area (F_{zr}) expressed in m^3/ha,
- capacity index (K), i.e., a parameter introduced by Annen & Londong (1966) and expressed as $K = V_{zb}/Q_{dmax}$,
- dimensionless capacity index (δ) being a function transforming design rainfall to dimensioning the design rainfall channel for determination of the tank volume. The function is dependent on the maximum flow reduction factor (β) and the coefficient of non-homogeneous outflow from the tank ($\chi = Q_{omin}/Q_{omax}$), where Q_{omin} – minimum flow from which stormwater retention in the tank begins.

In terms of the dimensions of the unit capacity parameters of the tank (W, K, δ), the methods for calculation of tank capacity can be divided into three groups. The first group includes methods with the dimensionless capacity index as the basis for determination of V_{zb} (Müller-Neuhaus 1953). The capacity factor (Annen & Londong 1966, Pecher 1972) and the unit capacity index (Błaszczyk 1974, Dziopak 1984, Mrowiec 2009) are used in the second and third groups, respectively.

7.3.1.5 Baker method

One of the simplest methods for determination of the required capacity of single-chamber and flow tanks is the Baker method, in which the tank inflow and outflow are described with the use of a triangular hydrograph (Figure 7.10 left), hence the following equation for calculation of tank capacity:

$$V_{zb} = \frac{1}{2} \cdot Q_{d\max} \cdot t_p \cdot (\lambda + 1) - \frac{1}{2} \cdot Q_{o\max} \cdot t_p \cdot (\lambda + 1) \tag{7.5}$$

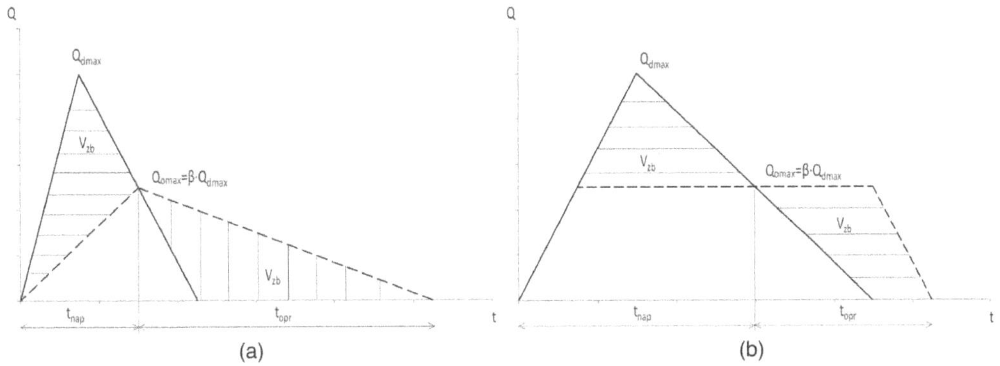

Figure 7.10 Calculation scheme for tank capacity in Baker (left), and Abt Grigg methods (right).

Appropriate transformations and substitution of Eq. 7.5 give:

$$V_{zb} = \frac{1}{2} \cdot t_p \cdot (\lambda + 1) \cdot Q(1 - \beta)_{d\max} \tag{7.6}$$

The following equation is derived by dividing both sides by V_c:

$$\frac{V_{zb}}{V_c} = 1 - \beta \tag{7.7}$$

In the Baker method, the dimensionless ratio (V_{zb}/V_c) is a linear function of the maximum flow reduction coefficient. Nevertheless, the results of laboratory analyses and computer simulations indicate that the tank outflow $Q_o = f(t)$ is a curvilinear function; hence, the relationship obtained diverges from reality.

7.3.1.6 *Abt and Grigg method with modifications*

Abt & Grigg (1978) considered the use of a triangular inflow hydrograph but assumed that the outflow hydrograph is a trapezoid (Figure 7.10 right). Hence, the required accumulation capacity of the tank can be defined as:

$$V_{zb} = \frac{1}{2} \cdot Q_{d\max} \cdot t_p \cdot (\lambda + 1) \cdot (1 - \beta)^2 \tag{7.8}$$

where the following relationship is obtained after appropriate transformations:

$$\frac{V_{zb}}{V_c} = (1 - \beta)^2 \tag{7.9}$$

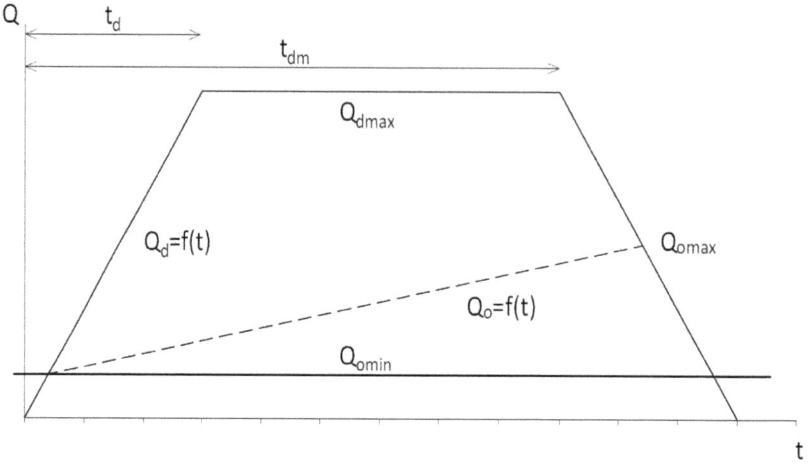

Figure 7.11 Calculation scheme for tank capacity in the Abt & Grigg method (1978).

As it is already known, the assumption of constant outflow from the tank may be associated with the use of a flow chamber; therefore, the Abt and Grigg method can be employed to estimate the capacity of multi-chamber tanks.

Since the maximum required accumulation capacity of the tank is determined by the hydrograph of rainfall with lower intensity and longer duration rather than the maximum instantaneous inflow (Dziopak 1997; Froehlich 2009), hydrographs of continuous rainfall should be considered in the design of the tank. Therefore, with many simplifications, mainly the assumption of constant rainfall intensity, the inflow hydrograph can be described by a trapezoid. Aron & Kibler (1990) and Guo (1999) transformed the rainfall-runoff phenomenon into a trapezoid and assumed a linear variability of the outflow from the tank (Figure 7.11); they expressed the tank capacity as follows:

$$V_{zb} = \frac{1}{2}\cdot(Q_{dmax} - Q_{omax})\cdot\left(t_{dm} + t_d\right) \tag{7.10}$$

where: Q_{dmax} – the maximum flow during the culmination defined as:

$$Q_{d\,\max} = q\cdot F_{zr} = \frac{a}{b+t_{\mathrm{dm}}}\cdot F_{zr} = \frac{K_1}{b+t_{\mathrm{dm}}} \tag{7.11}$$

where: F_{zr} – reduced catchment area calculated as $F\cdot\psi_{śr}$ (where: F – catchment area, $\psi_{śr}$ – mean runoff coefficient), a, b – regional parameters describing local rainfall characteristics.

The following equation is derived by substitution of (7.7) in (7.6) and appropriate transformations:

$$V_{\mathrm{zb}} = \frac{t_{\mathrm{dm}}\cdot K_1}{2\cdot\left(b+t_{\mathrm{dm}}\right)} + \frac{t_d\cdot K_1}{2\cdot\left(b+t_{\mathrm{dm}}\right)} - \frac{Q_{o\,\max}\cdot t_{\mathrm{dm}}}{2} - \frac{Q_{o\,\max}\cdot t_d}{2} \tag{7.12}$$

Knowing that the maximum accumulation capacity of the tank is obtained for the design t_{dm} value, if relation (7.12) reaches the extreme value and $dV_{zb}/dt_{dm} = 0$, the following equation can be proposed:

$$\frac{dV_{zb}}{dt_{dm}} = 0 = \frac{K_1 \cdot (b - t_d)}{(b + t_{dm})^2} - I_{o\max} \tag{7.13}$$

With appropriate transformations of (7.13), the design rainfall time for tank dimensioning can be calculated with the following formula:

$$t_{dm} = \sqrt{\frac{K_1 \cdot (b - t_d)}{Q_{o\max}}} - b \tag{7.14}$$

The required maximum tank capacity can be calculated after substitution of (7.14) in (7.12).

In the case of tubular tanks with an additional flow chamber, it can be assumed in some solutions (PROXIMUS) that the value of the outflow from the tank (Q_{omax} = const) is approximately constant (Figure 7.12) and therefore the required accumulation capacity in the trapezoidal inflow hydrograph is determined by the following equation:

$$V_{zb} = \left(t_{dm} - t_d \cdot \frac{Q_{o\max}}{Q_{d\max}} \right) \left(Q_{d\max} - Q_{o\max} \right) \tag{7.15}$$

Substitution of (Eq. 7.11) in (Eq. 7.15) and appropriate transformations give:

$$V_{zb} = \left(\frac{K_1 \cdot t_{dm}}{b + t_{dm}} - t_{dm} \cdot Q_{o\max} - t_d \cdot Q_{o\max} + \frac{Q_{o\max}^2 \cdot t_d \cdot (t_{dm} + b)}{K_1} \right) \tag{7.16}$$

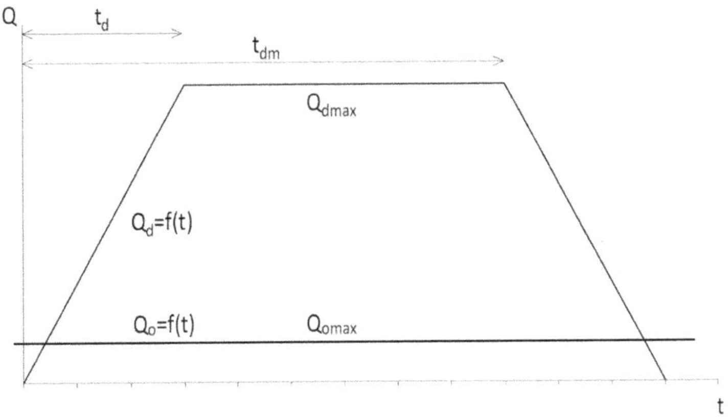

Figure 7.12 Scheme for calculating rainfall duration and reservoir capacity.

where: t_{dm}, b, Q_{omax}, t_d, a, b – as in (Eq. 7.11).

Knowing that the maximum accumulation capacity of the tank is obtained for the design t_{dm} value, if relation (7.16) reaches the extreme value and $dV_{zb}/dt_{dm} = 0$, the following equation can be proposed:

$$\frac{dV_{zb}}{dt_{dm}} = 0 = \left(\frac{K_1 \cdot b}{\left(t_{dm} + b\right)^2} - Q_{o\max} + \frac{Q_{o\max}^2 \cdot t_d}{K_1} \right) \tag{7.17}$$

With appropriate transformations of (Eq. 7.17), the design time in tank dimensioning can be calculated using the following formula:

$$t_{dm} = \sqrt{\frac{K_1 \cdot b}{Q_{o\max} - K_2}} - b \tag{7.18}$$

$$K_2 = \frac{Q_o^2 \cdot t_d}{K_1} \tag{7.19}$$

The required maximum tank capacity can be calculated after substitution of (7.18) in (7.16).

7.3.1.7 Simplified method for tank dimensioning according to DWA-A 117

The methodology for calculating tank design is discussed in the DWA-A 117 guidelines based on an appropriate calculation scheme depending on the catchment size. To determine the capacity of a tank located in a small (60–80 ha) paved urban catchment, a simplified dimensioning method is employed, provided that the frequency of capacity surcharge is not lower than $n = 0.1$/year and the outflow from the tank exceeds 2 l/(ha·s)$^{-1}$. In this method, the required accumulation capacity is calculated with the use of the following formula (DWA-A 117 2006):

$$V_{zb} = 0.06 \cdot D \cdot f_A \cdot f_Z \cdot \left(q_{D,n} - q_{\mathrm{Dr},R,u}\right) \cdot F_{zr} \tag{7.20}$$

where: f_A – reduction coefficient dependent on t_p (stormwater flow time), $q_{\mathrm{Dr},R,u}$, and n determined from appropriate nomograms (DWA-A 117 2006), $_r$ – impervious area of an adjacent catchment, $q_{\mathrm{Dr},R,u}$ $q_{D,n}$ – design rainfall intensity with duration D (min) and frequency n (l/(ha·s)$^{-1}$, $q_{\mathrm{Dr},R,u}$ – rainfall part of the throttled flow intensity on the impervious surface (A_u) determined as follows:

$$q_{\mathrm{Dr},R,u} = \frac{Q_{\mathrm{Dr}} - Q_{\mathrm{Dr},V} - Q_{T,d,\mathrm{aM}}}{A_u} \tag{7.21}$$

where: Q_{Dr} – intensity of throttled outflow from the tank (l/s), $Q_{\mathrm{Dr},V}$ – sum of outflow intensities from tanks located above the considered one (l/s), $Q_{T,d,\mathrm{aM}}$ – dry weather outflow intensity from the adjacent catchment (l/s).

Safety factor f_z assumed for low (1.20), medium (1.15), and high (1.10) (DWA-A 117 2006, 2006) risk or determined from an empirical relationship based on the logistic regression equation (own study):

$$f_z = 0.0267 \cdot \left\{ 5.33 + \ln\left(\frac{100 - R_s}{R_s} \right) \right\} \tag{7.22}$$

where: R_s – risk level, i.e., a parameter with values from 0 to 100.

Factor f_z in Eq.7.22 is an increase factor, which is determined depending on the level of risk. The risk function is a result of analyses of 10 different rainfall series from different parts of Germany for a surcharge frequency of 0.1–1.0/year and 40 $[\mathrm{l/(ha \cdot s)^{-1}}] > q_{\mathrm{Dr},R,u} > 2\ [\mathrm{l/(ha \cdot s)^{-1}}]$.

7.3.1.8 Pecher method with modifications

Another simplified approach to tank dimensioning according to Szeląg is the Pecher method (1972), in which a constant outflow from the tank (q_o) is assumed and pluviographic data are primarily used. The basis for calculation of the required accumulation capacity in this method is the equation of stormwater volume balance Eq. 7.4 transformed into the following form:

$$V_{\mathrm{zb}} = \int_0^{V_c} dV = \int_0^t Q_d(t)\,dt - \int_{t_o}^t dV_k - \int_{t_o}^t Q_o(t)\,dt = V_{\mathrm{inflow}} - V_{\mathrm{sew}} - V_{\mathrm{outflow}} \tag{7.23}$$

$$V_{\mathrm{inflow}} = F_{\mathrm{zr}} \cdot \left(P - h_{\mathrm{str}} - q_{\mathrm{par}} \cdot (t_d + t_k) \right) \tag{7.24}$$

$$V_{\mathrm{outflow}} = F_{\mathrm{zr}} \cdot q_o \cdot (t_d + t_k) \tag{7.25}$$

$$V_{\mathrm{sew}} = \sum_{i=1}^{m} h_i(t) \cdot L_i \tag{7.26}$$

where: V_{outflow} – volume of the tank outflow (m³), V_{inflow} – volume of surface runoff (m³), P – total rainfall depth (mm), q_{par} – evaporation losses expressed in dm³/s·ha (taken into account only for long-term rainfalls) and assumed as $q_{\mathrm{par}} = (1 \div 1.6)$ dm³/s·ha, h_{st} – initial loss volume assumed as 0.5 ÷ 2.5 mm, i.e., 5 ÷ 25 m³/ha, t_d – rainfall duration (min), t_k – field concentration time (min), V_{sew} – volume of retained stormwater.

The practical use of Eq. 7.24 was presented by Mrowiec (2009); however, in the analyses, the author did not include the accumulated stormwater volume. The graphic interpretation of Eqs. 7.23–7.25 proposed by Mrowiec (2009) is shown in Figure 7.13. It indicates that the calculation of tank capacity proceeds by shifting the tank characteristics along the ordinate axis and depends on the designer's decision. This decision is based on the statistics of permissible multiplicity of the overflow of the designed device. The diagram method and two modifications thereof are discussed in detail by Mrowiec (2009). This method can also be used for estimation of the frequency of switching on and operation time of pumps in gravity-pump tanks, as well as blowers in gravity-vacuum tanks. Since access to pluviographic data in catchments is limited,

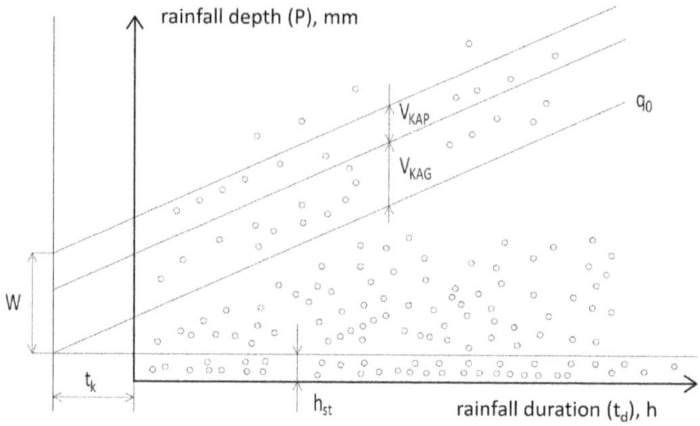

Figure 7.13 Example of the use of the diagram method to calculate the KAG and KAP capacity of a vacuum tank (Mrowiec 2009).

Notes: W – retention tank capacity, mm, V_{KAP} – vacuum accumulation chamber capacity, m³, V_{KAG} – gravity accumulation chamber capacity, m³.

the method is not applicable in uncontrolled urbanised catchments. Moreover, exclusion of the rainfall dynamics during the rainfall event is a substantial simplification, which may lead to errors in estimation of the capacity of the retention tank, as confirmed by the calculation results reported by Mrowiec (2009).

7.3.1.9 SCS curves

A simplified method for determination of the capacity of open-field reservoirs, taking into account the impact of rainfall intensity distribution in the case of any shape of the inflow hydrograph was presented by the American Soil Conservation Service. In this method, the required accumulation capacity relative to the total inflow hydrograph volume can be described by the following equations:

$$\text{SCS I} \quad \frac{V_{zb}}{V_c} = 0.56 - 6.39 \cdot \beta^2 + 13.77 \cdot \beta^3 - 8.33 \cdot \beta^4 \left(R^2 = 0.97\right) \tag{7.27}$$

$$\text{SCS II and III} \quad \frac{V_{zb}}{V_c} = 0.60 - 5.15 \cdot \beta^2 + 10.98 \cdot \beta^3 - 6.77 \cdot \beta^4 \left(R^2 = 0.99\right) \tag{7.28}$$

Investigations of single-chamber tanks (Hong 2008; Froehlich 2009) indicate that this method leads to underestimation of the tank capacity in terms of the maximum flow reduction factor $\beta = 0.1 \div 0.8$. This is also confirmed by the calculations for tubular tanks reported by Szeląg & Kiczko (2014).

7.3.1.10 Graphical and analytical method tubular tank dimensioning method

Since the relationship describing the outflow from a single-chamber tank differs largely from the linear one, which in turn leads to underestimation of the accumulation capacity manifested by, e.g., stormwater overflow, analytical-graphical methods have been proposed to increase the accuracy of calculations. Initially, they served for analysis of the operation of open-field reservoirs, as they ensured an optimal strategy for controlling discharges in flood events and facilitated the search for the maximum required area of flooding. In these cases, the calculation error may lead to substantial material losses, as shown in numerous studies (Dziopak 1984). These methods are based on the differential equation of the stormwater volume balance transformed into the form of finite differences. What is noteworthy is that their results are similar to those yielded by numerical methods. The method for calculation of the tank outflow hydrograph was discussed based on the Goodrich-Wisler approach (Orłowski 1977), in which Eq. 7.29 was transformed into the following form:

$$\left[Q_{d1}(t_1)+Q_{d2}(t_2)\right]+\left[\frac{2\cdot V_1(t_1)}{dt}-Q_{o1}(t_1)\right]=\frac{2\cdot V_2(t_2)}{\Delta t}+Q_{o2}(t_2) \tag{7.29}$$

where: $Q_{d1}(t_1)$, $Q_{d2}(t_2)$ – intensity of inflow to the tank at time points t_1 and t_2, respectively; $Q_{o1}(t_1)$, $Q_{o2}(t_2)$ – intensity of outflow from the tank at time points t_1 and t_2, respectively; $V_1(t_1)$, $V_2(t_2)$ – stormwater volume in the tank at time points t_1 and t_2 (m^3), respectively.

Graphs $G(Q_o) = 2\cdot V(H)/(dt)^{-1} \pm Q(H)$ (Figure 7.14) are prepared for the assumed tank capacity curve $V=f(H)$ and the characteristics of relief devices $Q=f(H)$. At the beginning of calculations, the inflow intensity $Q_d(t_1)$ is assumed to be equal to the outflow intensity $Q_o(t_1)$. Next, the value of $Q_d(t_1)$ is placed on the abscissa to the intersection with the curve $(2\cdot V/dt) - Q$. From the intersection point $G=f(Q_d(t_1))$, the sum of inflow intensities $Q_d(t_1)$ and $Q_d(t_2)$ is measured to the right-hand side and, at this distance, a line is drawn vertically to the curve intersection $(2\cdot V/dt)+Q$; $Q_o(t_2)$ is read on the abscissa.

Over the last 30 years, a number of graphical methods have been developed based on the numerical solution of the differential volume balance equation (Chow et al. 1998). The considerations were focused on open-field reservoirs, whose outflows had such relief devices as an overflow or drainpipe, or the curve of the flood surface was described with a power formula.

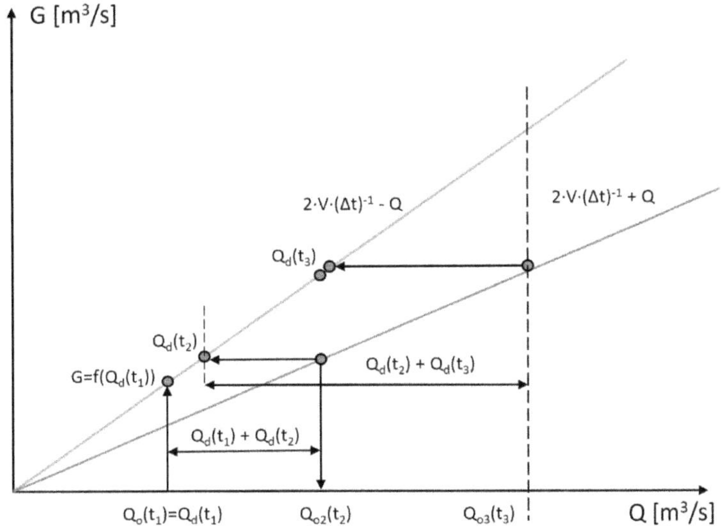

Figure 7.14 Calculation scheme in the Goodrich-Wisler method (Orłowski 1977).

Triangular, trapezoidal, curvilinear, and gamma-distribution inflow hydrographs were considered for calculation of the accumulation chamber. The basis for tank capacity calculations in these methods is the differential equation of the stormwater volume balance subjected to normalisation targeted at dependence of the tank capacity parameter on the hydraulic characteristics and the inflow hydrograph.

At present, the concept of a graphical method for dimensioning a tubular tank was presented by Szeląg & Kiczko (2014); however, since the prediction of the tank capacity was burdened with considerable error, this study presented modification of the method.

7.3.1.11 Analytical – graphical method for tubular tank

In the presented methodology, a triangular inflow hydrograph was assumed for tank dimensioning. As shown by (ATV A-118 2000; Hong 2008), in terms of the scope of this research, it is an appropriate representation of the response of a small urbanised catchment to short-duration high-intensity rainfall. The application of this type of parameterisation to describe the intensity of stormwater inflow and outflow as part of the tank-dimensioning task has been presented by, e.g., Guo (1999). The triangular inflow hydrograph (Figure 7.15) in the rise and recession phases can be described with the following equations:

$$Q(t) = \frac{Q_{d\max}}{t_p} \cdot t \tag{7.30}$$

$$Q(t) = Q_{d\max} \cdot \left(1 + \frac{1}{\lambda} - \frac{t}{\lambda \cdot t_p}\right) \tag{7.31}$$

where: $Q_{d\max}$ – peak discharge, t_p –rise time, λ – hydrograph fineness expressed as t_{op}/t_p.

Taking into account the hydrograph fineness (λ), the inflow hydrograph volume can be expressed as follows:

$$V_c = \frac{1}{2} \cdot Q_{\text{dmax}} \cdot t_p \cdot (\lambda + 1) \tag{7.32}$$

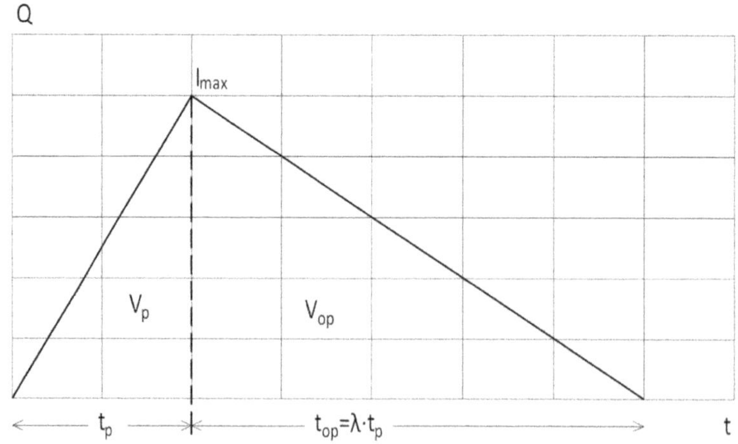

Figure 7.15 Inflow hydrograph scheme used for calculation of retention tank capacity.

where: Q_{dmax}, t_p, λ – as in Eqs. 7.30 and 7.31; Figure 7.15 illustrates these relationships.

The differential equation of the volume balance expressed in Eq. 7.4 was used to describe the relationship of the variability of the volume of stormwater accumulated in the tank $V(t)$ with the inflow $Q_d(t)$ and the outflow $O_o(t)$. There is relatively little information about the variability of the stormwater depth in the projection and the volume of stormwater accumulated in the tubular tank. This problem was discussed in detail by Mrowiec (2002), who proposed differential equations of stormwater volume balance for individual phases of tubular tank operation.

Assuming that the longitudinal slope of the accumulation chamber $i = 0$, the tank capacity $V(h)$ for any filling h can be expressed as follows:

$$V(h) = A(h) \cdot L_{zb} \tag{7.33}$$

where: L_{zb} – length of the accumulation chamber, $A(h)$ – wetted surface area (Figure 7.15) defined for the filling of the accumulation chamber h as follows:

$$A(h) = \frac{D_{ak}^2}{8} \cdot \left(2 \cdot \theta(h) - \sin 2 \cdot \theta(h)\right) \tag{7.34}$$

where: D_{ak} – diameter of the accumulation chamber (m), θ – central angle in radians corresponding to filling h equal to $\theta = \arccos(1–2 \cdot h/D_{ak})$.

Due to its complex form, Eq. 8.34 was approximated in the present analyses using the relationship proposed by Hager (2010):

$$\frac{A(h)}{D_{ak}^2} = \left(\frac{h}{D_{ak}}\right)^{1.4} \tag{7.35}$$

where: h – accumulation chamber filling (maximum filling of the tubular chamber is $h = h_m = D_{ak}$), $A(h)$ – cross-sectional area of the tubular tank.

By substitution of Eq. 7.35 into Eq. 7.33 and appropriate transformations, the tubular tank filling is calculated as follows:

$$h = \left(\frac{V(h)}{D_{ak}^{0.60} \cdot L_{zb}}\right)^{5/7} \tag{7.36}$$

where: L_{zb}, D_{ak}, $V(h)$ – as in equations (Eq. 7.33 and Eq. 7.35).

In the presented methodology, the outflow from the tubular tank through a circular drain orifice is represented as follows:

$$Q_0(h) = C_{sp}\sqrt{h} \tag{7.37}$$

where: C_{sp} – parameter of the characteristics of the drain orifice expressed as $C_{sp} = \mu \cdot \pi/4 \cdot D_{sp}^{2} \cdot (2 \cdot g)^{0.50}$ (where: μ – discharge coefficient, g – gravitational acceleration, D_{sp} – diameter of the circular drain orifice).

Taking into account the accumulation chamber filling in Eq. 7.36, the following relationship was obtained:

$$Q(h) = C_{sp} \cdot \left(\frac{V(h)}{C_z}\right)^{5/14} \tag{7.38}$$

where: $V(h)$ – capacity of the tubular tank at filling h, C_{sp} – parameter of the characteristics of the drain orifice, C_z – parameter calculated as $C_z = L_{zb} \cdot D_{ak}^{0,60}$.

The maximum flow rate through the drain orifice at complete filling of the accumulation chamber is expressed as

$$Q\left(h = D_{\text{ak}}\right) = C_{\text{sp}} \cdot \sqrt{D_{\text{ak}}} \tag{7.39}$$

hence the maximum flow reduction coefficient is equal to

$$\beta = \frac{Q\left(h = D_{\text{ak}}\right)}{I_{\text{max}}} \frac{C_{\text{sp}} \cdot \sqrt{D_{\text{ak}}}}{I_{\text{max}}} \tag{7.40}$$

where: C_{sp}, D_{ak}, Q_{dmax} – as in equations (Eqs. 7.31 and 7.37).

To solve the differential equation $V(h) = f(t)$, the explicit Runge-Kutta method was used in this study with regard to the length of the accumulation chamber L_{zb}. The model obtained in this way described the transformation of the inflow into the outflow, which is a function of the length of the tank L_{zb}. In the presented considerations, the problem of tank dimensioning has been reduced to identification of the value of a variable for which a specific rise reduction effect is obtained. The limitation consisted in the minimisation task performed with the use of the Nelder-Mead local optimisation algorithm (Nelder & Mead 1965). The minimised objective function had the following form:

$$F = \left(\beta_c - \beta_o\right)^2 \tag{7.41}$$

where: β_c – assumed value of reduction, β_o – value of reduction calculated by a numerical solution.

Therefore, the problem of tank dimensioning in these analyses was reduced to the following relationship:

$$L_{\text{opt}} = \arg\min F\left(L\right) \tag{7.42}$$

where: L_{opt} length of the tank at which the objective function reaches its minimum.

Due to the implicit form of the task in relation to the length, an additional condition describing the outflow from the tank in complete filling conditions was introduced in the presented methodology. It was based on an assumption that there was no retention in the network above the tank, resulting in equal inflow and outflow at $h = D_{ak}$. In the minimisation task, the inflow hydrograph parameters were the set values; hence, the analysis of the relationship between individual parameters required multiple performance of this task.

7.3.1.12 *Dimensionless differential equation of stormwater volume balance*

To limit the number of variables that should be taken into account while designing the tank and solving the differential equation of the stormwater volume balance (Froehlich 2009; Graber 2009), dimensionless variables linking the inflow hydrograph parameters with the characteristics of the designed tubular tank were introduced. This was achieved with the use of dimensionless state equations. Taking into account Eqs. 7.30, 7.31, and 7.38 in Eq. 7.4, the following differential equations describing the variability of the stormwater volume in the tubular tank in the rise and recession phases were obtained (Szeląg & Kiczko 2014)

$$\frac{dV}{dt} = Q_{d\max} \cdot \frac{t}{t_p} - C_{sp} \cdot \left(\frac{V}{C_z}\right)^{0.35} \tag{7.43}$$

$$\frac{dV}{dt} = Q_{d\max} \cdot \left(1 + \frac{1}{\lambda} - \frac{t}{\lambda \cdot t_p}\right) - C_{sp} \cdot \left(\frac{V}{C_z}\right)^{0.35} \tag{7.44}$$

where: $Q_{d\max}$, t_p, t, λ, C_{sp}, C_z – $Q_{\max}$ – maximum inflow; t_p – time of rising, λ – hydrograph fineness expressed as t_{op}/t_p, C_z – coefficient describing relation $L_z \cdot D_{ak}^{0.60}$, C_{sp} - parameter considered the characteristics of an orifice describing equation $C_{sp} = \frac{1}{4} \cdot \pi \cdot D_{sp}^2 \cdot \sqrt{2g}$.

Equations (9.43) and (9.44) shown above were transformed into a dimensionless form by introduction of the following parameters (Akan 1990, McEnroe 1992, Hong 2008):

- dimensionless time from the beginning (t^*)

$$t^* = \frac{t}{t_p} \tag{7.45}$$

- dimensionless inflow intensity (Q_d^*)

$$Q_d^* = \frac{Q}{Q_{d\max}} \tag{7.46}$$

- dimensionless tank volume (V^*)

$$V^* = \frac{V}{V_c} \tag{7.47}$$

where: t_p – rise phase time, V_c – inflow hydrograph volume, $Q_{d\max}$ – peak flow.

Substitution of Eqs. 7.45, 7.46, and 7.47 into Eqs. 7.43 and 7.44 yielded the following dimensionless equations (Szeląg & Kiczko 2014):

$$\frac{dV^*}{dt^*} = \omega \cdot t^* - \kappa_D \cdot \left(V^*\right)^{0.35} \tag{7.48}$$

$$\frac{dV^*}{dt^*} = \omega \cdot \left(1 + \frac{1}{\lambda} - \frac{t^*}{\lambda}\right) - \kappa_D \cdot \left(V^*\right)^{0.35} \tag{7.49}$$

where: κ_D – parameter taking into account the characteristics of the tank and the inflow hydrograph expressed as follows (Szeląg & Kiczko 2014):

$$\kappa_D = \frac{C_{sp} \cdot t_p}{C_z^{0.35} \cdot V_c^{0.65}} \tag{7.50}$$

where: ω – inflow hydrograph shape parameter defined with the formula:

$$\omega = \frac{Q_{d\max} \cdot t_p}{V_c} \tag{7.51}$$

In the cited paper, the volumes of the tubular tanks were calculated assuming that the maximum filling (h) is equal to the diameter of the accumulation chamber (D_{ak}). Since the tank volume (V) is a function of the tank length (L_{zb}), which depends on the value of the C_D parameter, the differential equations (Eqs. 7.43 and 7.44) were solved iteratively.

The calculation of the retention tank volume in these considerations was based on triangular inflow hydrographs with volumes $V_c = 900 \div 45000\,m^3$, maximum flow rate $Q_{dmax} = 1.5 \div 7.5\,m^3/s$, and rise time $t_p = 10 \div 150$ minute. The values of the analysed hydrograph shape parameters ω ranged from 0.15 to 1.00. The analyses were performed for $D_{sp} = 0.25 \div 1.25$ m orifice diameter and $D_{ak} = 1.25 \div 6.00$ m accumulation chamber diameter. The calculations of the tubular tank volumes were conducted for reduction coefficient values of $\beta = 0.1 \div 0.9$. The analysis of such a large range of the variability of the inflow hydrograph parameters and sizes of the tubular tank required over 1,000 simulations, as shown in the study. The calculations were used for preparation of a nomogram (Figure 7.16) showing the impact of the hydrograph shape parameter (ω) on the κ_D parameter for constant values of the maximum flow reduction coefficient ($\beta = 0.1 \div 0.9$).

On the basis of Figure 7.16, it was concluded that relation $\kappa_D = f(\omega, \beta = \text{const})$ is the following linear function:

$$\kappa_D = \xi \cdot \omega + \varepsilon \tag{7.52}$$

where: ξ, ε – empirical parameters characteristic for each value of the maximum flow reduction coefficient determined with the highest probability method.

These calculations indicate that, at extrapolation of the variability (Figure 7.16) with a linear function, the values of the coefficients of determination (r^2) in Eq. 7.52 for $\beta = 0.1 \div 0.9$ vary from 0.989 to 0.996. Moreover, based on the $\kappa_D = f(\omega, \beta)$ nomogram generated in the study (Figure 7.16), it was found that the higher the value of the hydrograph shape parameter (ω), the greater the κ_D parameter, assuming that β = const. For instance, the κ_D value is 0.135 for $\beta = 0.40$ and $\omega = 0.30$ and 0.390 for $\omega = 0.90$. Furthermore, the higher the maximum flow reduction coefficient, the higher the κ_D values. For example, at $\omega = 0.60$, the κ_D values for $\beta = 0.40$ and $\beta = 0.60$ are 0.268 and 0.483, respectively. The nomogram generated in the study indicates that the greater the value of the maximum flow reduction coefficient and the inflow hydrograph asymmetry factor, the greater the dispersion of the κ_D parameter values for the analysed hydrograph asymmetry factors ($\omega = 0.2 \div 1.0$), and thus the greater the error of the tank capacity prediction.

Given these considerations, it is advisable to perform a detailed analysis to determine the impact of the inflow hydrograph parameters and the hydraulic characteristics of the orifice and the accumulation chamber on the error in tank capacity prediction. Moreover, in order to increase the accuracy of the nomograms presented in the study, it is necessary to predict the size of this error. To this end, the multiple regression method, artificial neural networks, and the genetic programming method were used in the study.

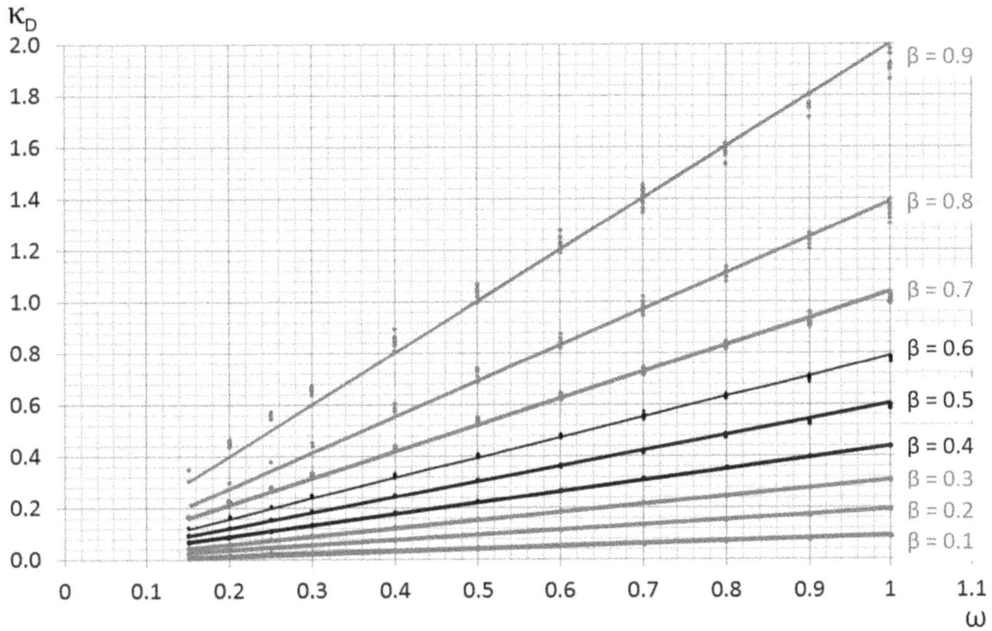

Figure 7.16 Effect of the κ_D parameter and the hydrograph asymmetry factor (ω) on the maximum flow reduction coefficient (β); the points correspond to simulation-derived values, and the curves show values obtained from Eq. 7.52 (Based on Szeląg et al. 2021b).

7.3.1.13 Analysis of the impact of the inflow hydrograph parameters and tank characteristics on the error in tank capacity prediction

The results of the calculations were used for determination of the theoretical values of the error of the accumulation chamber length based on the $\kappa_D = f(\omega, \beta)$ nomogram. Next, regression models were developed to calculate the error value based on the inflow hydrograph parameters and the characteristics of the accumulation chamber and the orifice. To this end, multiple stepwise regression, artificial neural networks, and the genetic programming method were used. Since there are no specific guidelines on the selection of the optimal network architecture for identification of the best neural networks, the Automatic Designer option in the STATISTICA program was used in the study. Linear, hyperbolic, sine, hyperbolic tangent, exponential, and logistic activation functions were considered for the neurons of the hidden and output layers. To carry out the learning process properly and to evaluate the operation adequately, the theoretical values of the error in the prediction of the accumulation chamber length were divided into three sets (training – 50%, validation – 25%, and test – 25%). The process of neural network learning was carried out with the Broyden-Fletcher-Goldfarb-Shanno method (Rutkowski 2006. To select an ANN structure that is the best in prediction of error ε, 100 artificial neural networks were generated at the stage of calculations.

To determine the impact of the $\kappa_D = f(\omega, \beta)$ approximation on the error in the tank capacity prediction obtained from numerical calculations, relative errors were determined with the formula:

$$\varepsilon_V = \frac{V_{zb[sym]} - V_{zb[nom]}}{V_{zb[sym]}} = \frac{F_{ak} \cdot \left(L_{zb[sym]} - L_{zb[nom]}\right)}{F_{ak} \cdot L_{zb[sym]}} = \frac{\Delta L_{zb}}{L_{zb[sym]}} \tag{7.53}$$

where: $V_{zb[Sym]}$ – tank capacity determined by numerical simulations (m³), $V_{zb[nom]}$ – tank capacity determined from generated nomograms (m³), $L_{zb[sym]}$ – tank length determined by simulations (m), $L_{zb[nom]}$ – tank length determined from the generated nomogram (m), F_{ak} – cross-sectional area of the accumulation chamber (m²), ΔL_{zb} – absolute error of tank length prediction that can be determined from the formula:

$$\Delta L_{zb} = \left(\frac{C_{sp} \cdot t_p}{D_{ak}^{0.21} \cdot V_c^{0.65}} \right)^{2.857} \cdot \left(\frac{1}{\kappa_{sym}^{2.857}} - \frac{1}{(\alpha \cdot \omega)^{2.857}} \right) \tag{7.54}$$

where: κ_{sym} – parameter determined by numerical calculations.

The values of relative errors (ε_V) that appear while applying the graph were determined from formula (7.53) based on the generated nomogram $\kappa_D = f(\omega, \beta)$ (Figure 7.16) and numerical calculations. Table 7.1 shows only the maximum ε_V.

As indicated by the calculations (Table 7.1), when the nomogram $\kappa_D = f(\omega, \beta)$ is used for determination of tank capacity, the smallest maximum relative error (ε_{Vmax}) is noted for $\beta = 0.1$ and its extreme values are in the range of $-7.01 \div 3.77$. In turn, the greatest error, ranging from −36.33% to 22.13% is obtained at $\beta = 0.9$. This indicates the significance of the values of relative errors in tank capacity prediction, which proves the need for prediction thereof. First, the multiple stepwise regression method was used for the determination of ε_V, as variables with a negligible impact on the relative error of the accumulation chamber capacity were eliminated at the stage of calculations. Thus, the following equation was obtained:

$$\varepsilon_V = 0.788(\pm 0.039) \cdot \kappa_D + 0.714(\pm 0.041) \cdot \omega + 1.621(\pm 0.086) \cdot Q_{d\max} + 0.200(\pm 0.002) \cdot t_p - 2.614(\pm 0.073) \cdot D_{sp} - 0.852(\pm 0.079) \cdot D_{ak} \tag{7.55}$$

The variables from Eq. 7.55 were used for construction of prognostic models based on artificial neural networks and genetic programming, i.e., the κ_D parameter, the coefficient of stormwater volume distribution unevenness in the hydrograph (ω), the peak flow during culmination (Q_{dmax}), the rise time (t_p), and the diameters of the outlet orifice and accumulation chamber. The parameters of the mathematical models proposed in the study are shown in Table 7.2, while Figure 7.17 presents a comparison of the ε_V values, which were both calculated and determined based on numerical simulations and the generated nomogram $\kappa_D = f(\omega, \beta)$. The calculations (Table 7.2) indicated a comparable agreement of the calculated and measured values of relative errors (ε_V) obtained with the use of the artificial neural network with 6 neurons in the hidden layer and exponential activation functions, and with the use of genetic programming. This was confirmed by the values of certain fitting parameters (Table 7.2), as the correlation coefficient

Table 7.1 Maximum values of relative errors (εVmax) for maximum flow reduction coefficients $\beta = 0.1 \div 0.9$

β	0.1	0.2	0.3	0.4	0.5	0.6	0.7	0.8	0.9
ε_{Vmax}	3.77	5.75	6.16	6.31	9.45	11.19	12.92	18.44	22.13
ε_{Vmax}	−7.01	−7.28	−9.79	−12.07	−18.08	−20.51	−22.93	−25.03	−36.33

Table 7.2 Measures of the goodness-of-fit of models predicting the relative error of tank capacity (ε_V)

Method	*WBR*	*MPE*	*r*	*AIC*
regression	0.0442	47.51	0.6851	−3553
ANN	0.0272	5.60	0.9310	−3010
GP(ω, κ_D)	0.0345	8.02	0.8872	−3321
GP(ω, β)	0.0312	7.89	0.8953	−3299
GP(ω, β)	0.0296	7.21	0.9010	−3236

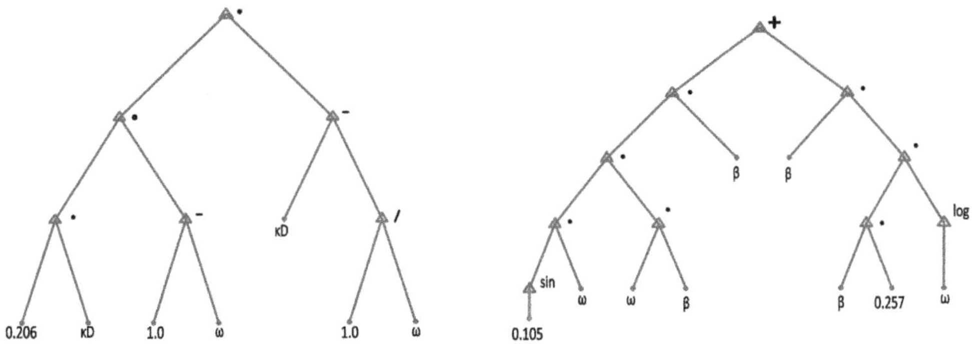

Figure 7.17 Structures of parsing trees for calculation of ε_V = (ω, κ_D, β) (Szeląg et al. 2021b).

and Akaike information criterion in the case of multiple regression as well as ANN and GP(ω, β) were 0.785 and −3553, 0.931 and -3010, and 0.901 and – 3236, respectively.

The differences in the fit parameters of the generated models to the theoretical data in the case of ANN and GP are insignificant. The correlation coefficient is only 3% higher and the average percentage error is 1.61% lower. The method of genetic programming yields the following dependencies for determination of the relative error of the capacity prediction:

$$\varepsilon_V = 0.2036 \cdot \kappa_D \cdot (1-\omega) \cdot (\omega^{-1} - \kappa_D) \quad (r = 0.8872) \tag{7.56}$$

$$\varepsilon_V = (\omega \cdot \beta)^2 \cdot \sin(0.105) + 0.2570 \cdot \beta^2 \cdot \ln(\omega) \quad (r = 0.8953) \tag{7.57}$$

$$\varepsilon_V = 0.1050 \cdot (\omega - 0.7917) \cdot \ln(0.625 \cdot \omega) - 0.376 \cdot \beta \quad (r = 0.9010) \tag{7.58}$$

which were determined based on the obtained parsing tree structures (Figures 7.17 and 7.18).

On the basis of Figures 7.19–7.21, it was found that the artificial neural networks were more accurate in prediction of relative error values than multiple regression and regression dependence derived from genetic programming calculations for $\varepsilon_{V[\text{sym}]} < -0.10$.

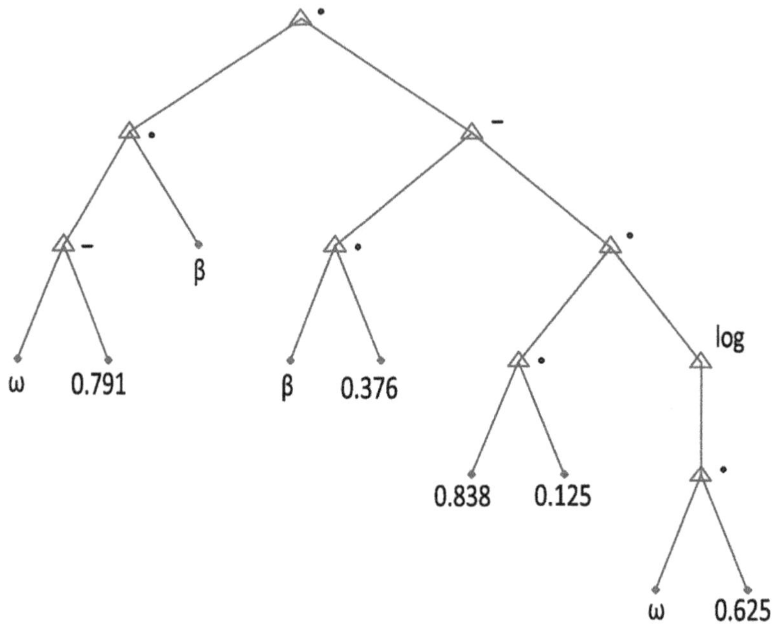

Figure 7.18 Structure of parsing trees for calculation of $\varepsilon_V = (\omega, \beta)$ (Szeląg et al. 2021b).

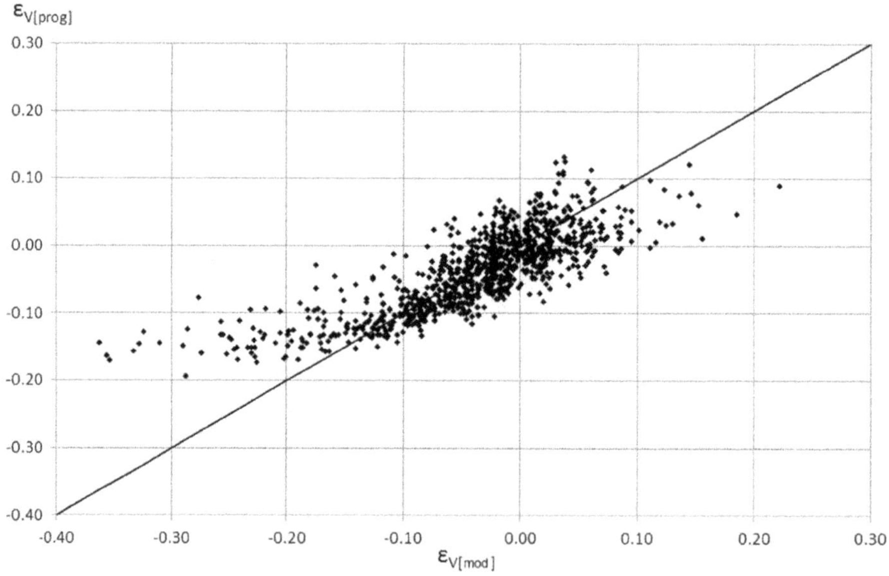

Figure 7.19 Comparison of error values obtained from the developed nomograms $\varepsilon_{V[sym]}$ and calculated using multiple regression $\varepsilon_{V[prog]}$ (Szeląg et al. 2021b).

7.3.1.14 Analytical – graphical method for rectangular tank

7.3.1.14.1 STOCHASTIC METHODS OF TANK DIMENSIONING

Stochastic methods for tank sizing take into account the random nature of rainfall, which can be modelled with various statistical methods applied only to long-term rainfall measurement results (DWA-A 117 2006). Analytical-probabilistic models are one of the dynamically developing

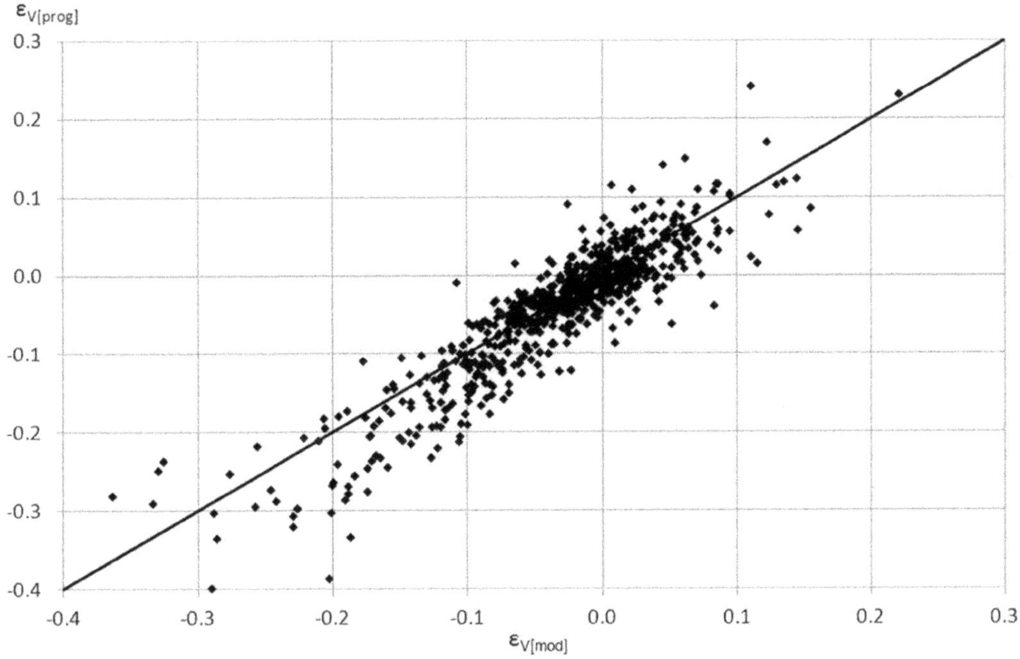

Figure 7.20 Comparison of error values obtained from the developed nomograms $\varepsilon_{V[sym]}$ and calculated using genetic programming $\varepsilon_{V[prog]}$ (Szeląg et al. 2021b).

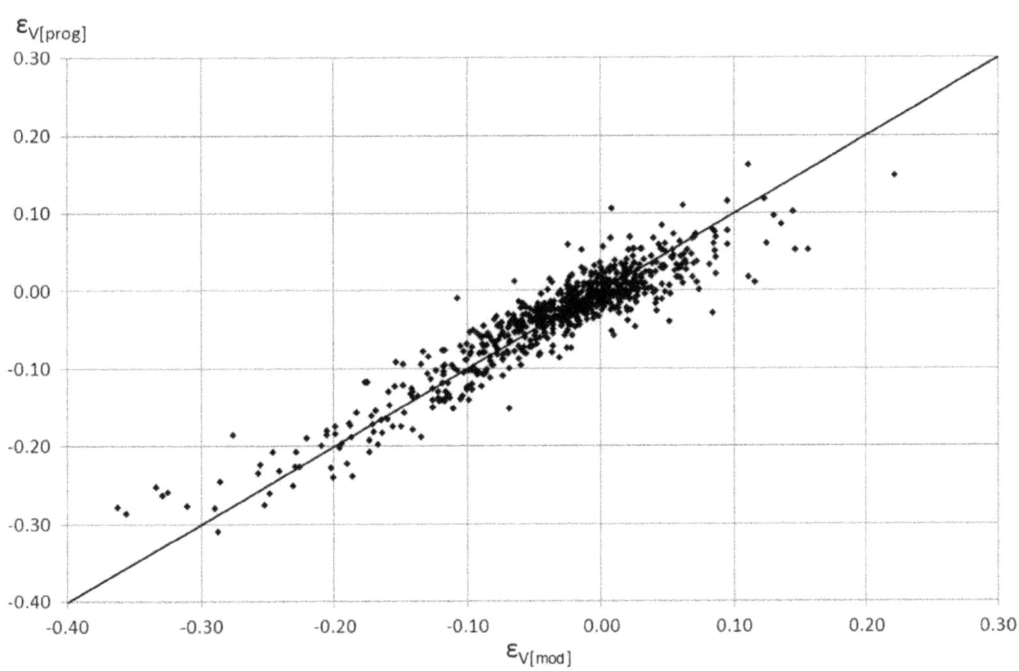

Figure 7.21 Comparison of error values obtained from the developed nomograms $\varepsilon_{V[sym]}$ and calculated using ANN $\varepsilon_{V[prog]}$ (Szeląg et al. 2021b).

stochastic methods. They schematise the outflow hydrograph with the use of a rectangle, assume constant rainfall intensity, and ignore or simplify the phenomenon of stormwater infiltration into the ground. It is also assumed that the rainfall parameters, i.e., the total amount and duration of rainfall, as well as the dry period are statistically non-correlated (Papa et al. 1999). In practice, this allows using the relationships between random distributions of two functionally related variables and generating equations for the probability of tank capacity surcharge. Synthetic rainfall generators predicting the distribution of rainfall intensity, as well as rainfall duration and depth may also be a basis for tank dimensioning. Multiyear rainfall measurement results are used to create these generators. Next, using copula functions, multidimensional probability density distributions are generated (Vandenberghe et al. 2010; Vernieuwe et al. 2015) for simulation of long-term rainfall observation series. The Monte Carlo (MC) method is used for prediction of rainfall from determined statistical distributions. It is employed to solve problems using random numbers with a specific distribution as samples describing the process. Therefore, the MC method is based on numerical calculations, where the analysed process has a stochastic character and the numerical solution is replaced by a probability task. The Monte Carlo method is based on manifold selection (sampling) of the value of variable X (j – identified hydrodynamic model parameters) in a series of independent trials. With a sufficiently large number of simulations, it yields an arithmetic mean of modelled values. In extreme cases, when the variables are independent, i.e., not correlated, it is possible to omit the creation of multidimensional distributions, which requires the use of complex statistical algorithms. An alternative solution is the use of multifractal geometry and modelling of rainfall series using canonical microcascades (Licznar et al. 2011). The results of rainfall simulations obtained with the use of rainfall generators can be used for simulations in hydrodynamic and conceptual models or other methods.

Over the last 30 years, more focus has been placed on the analysis of the uncertainty of estimated parameters in hydrodynamic catchment models, which is used for identification of catchment parameters (roughness coefficient, catchment retention, flow path width, impervious area), sewer network parameters (absolute roughness), and parameters of the dynamics of accumulation and removal of contaminants from the catchment. Analyses carried out by many researchers (Shrestha et al. 2009; Szeląg et al. 2014; 2016; 2019) demonstrated that it is not possible to establish one set of parameters of catchment characteristics in the physical model for which a satisfactory solution will be obtained, but there are many such solutions. Therefore, at the stage of calculation of tank capacity, it is not possible to determine an unequivocal inflow hydrograph. Many such calculations must be performed with many different combinations of parameter values from an appropriate range for which satisfactory agreement of calculated and measured hydrograph results has been obtained. This is important for practical considerations, as the inclusion of uncertainty in the calculation of tank capacity may lead to identification of events resulting in tank overfilling. Moreover, in the case of the developed hydrodynamic catchment models, this may facilitate identification of tank capacity ensuring failure-free and safe operation of a network comprising the tank.

The practical considerations in the analysis of the impact of uncertainty in estimation of the parameters of the hydrodynamic model (importantly, it must be calibrated) on the tank retention capacity can be reduced to the following calculations:

a determination of the ranges of hydrodynamic model parameters with satisfactory fitting results of:

- an outflow hydrograph measured and calculated in the model,

- tank filling measured and obtained from the simulation,
- in an extreme case, an inflow hydrograph measured and obtained from calculations,

b Monte Carlo simulation based on assumed statistical distributions of model parameters,
c hydrodynamic simulations and determination of the tank capacity,
d determination of the distributions of tank capacity variability using the distribution function.

7.3.1.15 Probabilistic method for determination of design rainfall duration for dimensioning the overflow tank

The problem of designing the optimal structure and required capacity of a retention tank is an important issue addressed by many researchers. Overflow tanks ensuring constant outflow intensity, which is their advantage over other solutions, are one of the commonly used structures of retention tanks in sewer systems. Therefore, a number of theoretical studies (Freni et al. 2008; De Martino et al. 2011; De Paola & De Martino 2013; Licznar 2013; Szeląg & Kiczko 2014;) have been undertaken to develop engineering methods for estimation of their capacity. In these considerations, the establishment of the tank capacity consists in determination of the duration of rainfall (so-called design rainfall) for which the calculated capacity is a maximum value for the assumed tank outflow and the standard of land drainage depending on local conditions (ATV-A 118 1999). A number of assumptions are made to determine the design rainfall duration, e.g., simplification of the surface runoff model and exclusion of the probabilistic nature of rainfall events. The capacities calculated in this way tend to be highly inaccurate (Dziopak & Słyś 2007; Paik 2008; De Paola & De Martino 2013); hence, calibrated hydrodynamic models are used to determine their size.

The basis for determination of the tank capacity in this approach is provided by the results of the so-called continuous simulations made with the use of a hydrodynamic catchment model based on multiyear rainfall observation series (30 years). In the absence of such series, synthetic rainfall generators are created. In the absence of data on rainfall, regional rainfall models are used in engineering considerations, which serve for determination of the maximum rainfall depth for the assumed rainfall duration and frequency (so-called IDF curves, intensity-duration-frequency) based on established empirical relationships. This approach is commonly used at the stage of modelling facilities located in sewer networks. An aspect that is frequently ignored at the stage of modelling the tank capacity is the uncertainty of identification of coefficients in the hydrodynamic model. This is puzzling, as the problem has been discussed in many studies (Andrés-Doménech et al. 2010; Szeląg 2013; Strzec et al. 2018; 2019; 2021) focused on simulation of the quantity and quality of stormwater runoff from the catchment. Simulation calculations have revealed a considerable impact of the uncertainty of hydrodynamic catchment models on the simulation results, which may result in wrong decisions on sewer system management. Therefore, an original method of retention tank dimensioning taking into account the uncertainty of the calibrated hydrodynamic model in relation to the surface runoff is presented below.

In the absence of continuous multiyear rainfall data, the retention tank capacity is determined from the frequency of computational rainfall – model rainfall of return period (C = 1, 2, 5, 10 years) established based on the category of the land drainage standard (PN-EN 752: 2008, ATV-A 118 1999). The calculation scheme of this method is shown in Figure 7.22.

The computational data include the geographic and physical characteristics of the catchment area, as well as rainfall and flow measurements used to calibrate the hydrodynamic model.

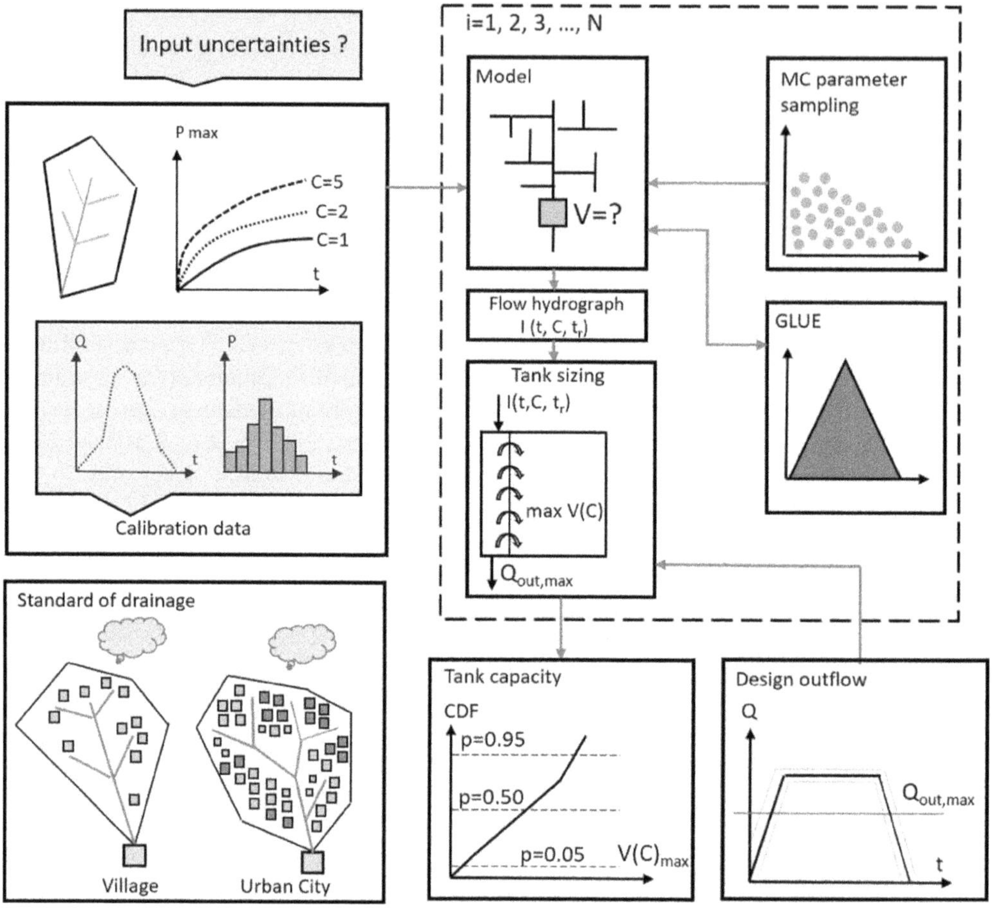

Figure 7.22 Calculation scheme of retention tank dimensioning (Based on Szeląg et al. 2019).

To determine the tank retention capacity, it is necessary to transform the rainfall depth value calculated using formulas (4.34 and 4.35) into the outflow value. To take into account the uncertainty of the catchment model, the Monte Carlo method is used for simulation of model parameters for the assumed ranges of their variability. Next, based on the measurements, the distributions of the analysed parameters are identified with the GLUE method. This is needed for determination of catchment outflow hydrographs, which together with the assumed tank structure, are used for calculation of the required tank capacity with the volume balance equation described by formula (7.4). In order to determine the maximum tank capacity for the assumed land drainage standard (C) and the assumed outflow (Q_o), it is necessary to perform calculations for different rainfall durations (t_r) to identify the extreme variability $V = f(C, t_r, Q_o)$ and the design rainfall duration, where $V \rightarrow \max\{(C, t_r, Q_o)\}$. The values of the retention tank outflow (Q_o) and the determined retention capacities (V) are normalised by relating the values of the variables mentioned above to the impervious catchment area (F_{imp}) and calculating the unit capacity (V_q) and unit outflow (q) coefficients.

Table 7.3 Ranges of variability of calibrated coefficients assumed for GLUE calculations

Parameters	*Units*	*Range*
Flow path width coefficient (a)	–	2.7–4.7
Retention of impervious areas (d_{imp})	mm	0.8–4.8
Retention of pervious areas (d_{perv})	mm	0.8–6.8
Roughness coefficient of impervious areas (n_{imp})	$m^{-1/3}·s$	0.01–0.022
Roughness coefficient of pervious areas (n_{perv})	$m^{-1/3}·s$	0.16–0.20
Channel roughness coefficient (n_n)	$m^{-1/3}·s$	0.01–0.048
Correction factor for the proportion of impervious area in the catchment (θ)	–	0.7–1.275
Correction factor for the mean subcatchment slope (μ)	–	0.8–1.375

Following the methodology presented in Figure 7.22, the ranges of the variability of individual model parameters were determined based on the results of SWMM calculations and literature data (Table 7.3). Due to the strong interactions between the calibrated parameters and their high impact on the results of surface runoff calculations, modification factors were introduced at the stage of calculations. They included the flow path width (W_j) coefficient (α), coefficients of the retention of impervious and pervious areas (d_{imp}, d_{perv}), roughness coefficients for impervious and pervious areas (n_{imp}, n_{perv}), coefficient (θ) for the proportion of impervious catchment areas (Imp_j) in the total catchment area, and coefficient (μ) for the mean subcatchment slope (J_j). To reduce the multidimensionality of the solution and limit the calculations, it was assumed that the flow path width was determined as $W = \alpha \cdot A_j^{0.5}$. The values of coefficients (μ, θ) were used to correct the value for the impervious area (Imp_j) and catchment slope ($J = \mu \cdot J_j$), where j is the number of subcatchments.

On the basis of uniform *a priori* distributions of the model parameters, 5,000 simulations of catchment outflow hydrographs were performed with the Monte Carlo method to determine their variability. The simulations were used for determination of *a posteriori* distributions of these parameters, which was one of the stages of the research (Figure 7.22) discussed in detail by Kiczko et al. (2018). The Monte Carlo sample was conditioned using a likelihood function on two sets of observations: for the events on 15 September 2010 and 7 August 2011. The scaling factor κ was adjusted to ensure that 95% of observations in calibration sets would be enclosed in 95% confidence intervals (Kiczko et al. 2018, Szeląg et al. 2016).

Figure 7.23 presents verification of the probabilistic solution for an independent data set. The simulations indicate high goodness-of-fit of the measured and modelled outflow hydrographs for the calibration set in the case of high-intensity rainfalls (time and total rainfall depth of $t_r = 270$ minute and P_{tot}=16.5 mm, which gives rainfall intensity of $q = 10.19$ $dm^3/(ha \cdot s)^{-1}$ at a hydrograph volume of 3,415 m^3).

Using the Monte Carlo simulation results and the *a posteriori* distributions of the model parameters (Table 7.3), catchment outflow hydrographs were calculated for rainfall $t_r = 10 \div 240$ minute and $C = 1, 2, 5, 10$ determined from the IDF curves. Next, based on Eq. 7.4, the capacities of retention tanks were determined for the assumed design flows ($Q_{out,max} = 0.1 \div 1.0\,m^3/s$), thus obtaining a probabilistic solution and determining the mean value and the 95% confidence interval for the individual quantities $V(t_{td},C,q)$. The results were compared with tank capacities calculated with the assumption that the values of parameters in the hydrodynamic model (SWMM) are the only set of parameters yielding satisfactory simulation results. This approach

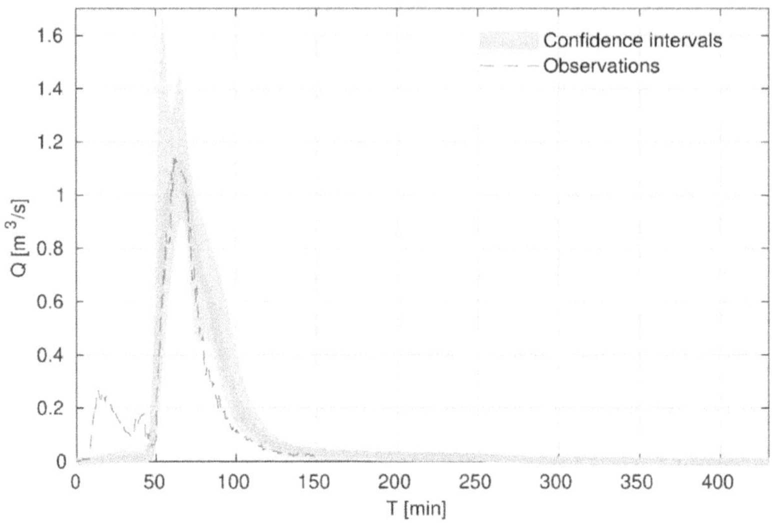

Figure 7.23 Examples of measurement and calculation results of the catchment outflow hydrograph for the 95% confidence interval obtained with the GLUE + GSA method (07.30.2010).

is commonly used in practice and is referred to as the deterministic solution. The calculation results facilitated assessment of the effect of the uncertainty of calibrated SWMM parameters on the determined tank capacities.

Given these remarks and the SWMM simulations, the impact of the uncertainty of the hydrodynamic catchment model on the design rainfall duration used for tank dimensioning and tank unit capacity was analysed. On the basis of the calculations, the $V_q = f(t_r)$ relationship was first determined for selected q values. An example of the $V_q = f(t_r)$ relationship for $C = 5$ as well as $q = 17$ dm^3/(ha·s) and $q = 6$ dm^3/(ha·s)$^{-1}$ is shown in Figure 7.24. The results allow drawing the conclusion that the uncertainty of the hydrodynamic catchment model exerts an impact on the unit retention capacity of the tank, as confirmed by the variability of the 95% confidence interval for individual $V_q = f(t_r)$ values and the $V_q = f(q,C)$ relationship shown in Figure 7.25. Moreover, it was found that the tank capacity determined taking into account the uncertainty (median) was lower than the value yielded by the deterministic solution, which is also indicated by the variability of the V_q value illustrated in Figure 7.24. Since the curves in Figure 7.24 were prepared for the selected q values, it is impossible to generalise them for the other unit values of tank outflows. To this end, it was necessary to analyse the relationship of the unit tank capacities obtained for t_{rm}^{prob} and t_{rm}^{det} with the assumed outflows (q) and the assumed rainfall frequency (C), which is discussed later in the paper.

On the basis of the curves determined for $q = 6$ dm^3/(ha·s) and q = 17 dm^3/(ha·s) shown in Figure 7.24 it is possible to determine the effect of the SWMM uncertainty on the determined design rainfall duration used for tank dimensioning with deterministic and probabilistic solutions. The curves indicate that the value of the design rainfall duration in the deterministic solution is shorter than that provided by the probabilistic solution (median).

In the deterministic solution (t_{rm}^{det}) for $q = 17$ dm^3/(ha·s), the t_{rm} value is 56 minute, which is by 19 minute shorter than in the probabilistic solution, and the ratio of $t_{rm}^{prob}/t_{rm}^{det} = 1.33$. In turn,

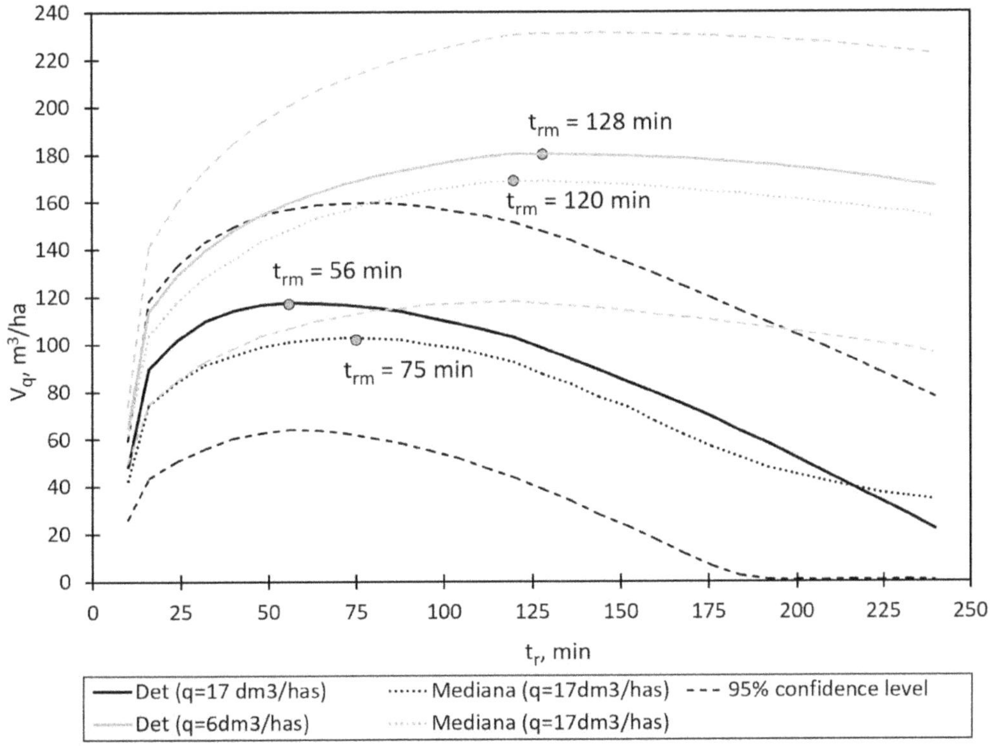

Figure 7.24 Impact of the unit tank outflow (q) and rainfall frequency (C) on the unit accumulation capacity (V_q) taking into account uncertainty of the model.

for $q = 7$ dm³/(ha·s), the shift between the rainfall durations in the considered solutions is shortened to 8 minute and the ratio $t_{rm}^{prob}/t_{rm}^{det} = 1.09$ is reduced correspondingly.

This result shows that the smaller the unit outflow from the reservoir (q) is, the smaller the value of the shift between the rainfall durations in deterministic and probabilistic solutions is, which determines the size of the required retention tank capacity.

In the next stage of the analyses, the calculations performed in the SWMM program were used for generation of curves (Figure 7.25) showing the effect of the tank outflow (q) on the unit maximum accumulation capacity (V_q) at the assumed C values for the design rainfall durations (t_{rm}). The calculations showed that the value of unit accumulation capacity (V_q) provided by the deterministic solution was higher than the median obtained in the probabilistic solution and decreased with the increase in the unit tank outflow, which was also confirmed by the $V_q = f(t_r, C)$ variability in Figure 7.24. This result indicates that the unit retention capacity of the tank (V_q) determined by the deterministic solution is overestimated in comparison with the capacity shown by the probabilistic solution (median), which means that the tank may be oversized. This is a favourable solution from the point of view of the reliability of the tank operation. However, given the determined 95% confidence interval, it is difficult to formulate far-reaching generalisations, as it may turn out that the tank capacity designed using the deterministic solution may still be insufficient. Therefore, further analyses are required in this regard, as retention

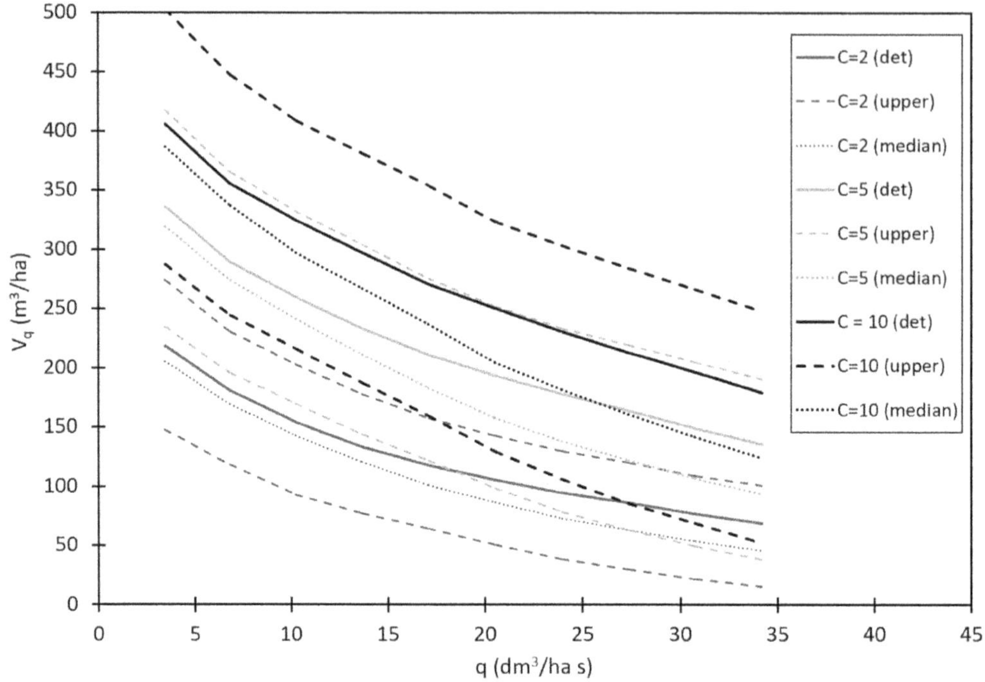

Figure 7.25 Influence of the unit tank outflow (q) and rainfall frequency (C) on the unit accumulation capacity (V_q) taking into account uncertainty of the model (Based on Szeląg et al. 2019).

tanks are essential objects in sewer networks, and the appropriate selection of their dimensions is highly important for the reliability of the operation of the drainage system.

The analysis of the obtained $V_q = f(q,C)$ curves indicates an increase in the relative difference in the unit tank capacities determined with the deterministic and probabilistic solutions depending on the q value. For instance, for $C = 2$, an increase from $q = 6.8$ dm³/(ha·s) to $q = 13.7$ dm³/(ha·s) results in an increase in the relative difference in tank capacity from 6% to 19%. A comparison of these results with the aforementioned analyses allows drawing a conclusion that an increase in the tank outflow volume is accompanied by an increase in the degree of tank oversizing (taking the probabilistic solution as a starting point). This is unfavourable for the economic aspect and the operation of the sewer system below the tank.

On the basis of the data shown in Figure 7.25, it can be concluded that an increase in the unit outflow from the tank (q) leads to a reduction in the required unit accumulation capacity (V_q). This implies that the uncertainty of the model is important for the selection of discharge devices. A comparison of the values of unit tank outflows for the assumed unit accumulation capacity allows drawing a conclusion that the q values are higher in the deterministic solution than when uncertainty is taken into account. This is of key importance for the tank design, as exclusion of uncertainty leads to overestimation of the drain outflow, which may result in unfavourable phenomena (system overload) in the stormwater network below the tank outlet.

As indicated by the calculations presented above, the uncertainty in the model has a significant impact on the retention tank design. The simulations showed that, due to the increase in the unit tank outflow to $q = 20$ dm³/(ha·s)$^{-1}$, the design rainfall duration taken into account in tank dimensioning (probabilistic solution) was longer than that in the deterministic solution. Simultaneously, with the increase in q, the difference in the values of design rainfall durations determined in the probabilistic and deterministic solutions increased. In turn, after exceeding a certain threshold, a further increase in q led to a decrease in the difference between the rainfall durations provided by the probabilistic and deterministic solutions.

These analyses indicated that, in the case under consideration, the unit accumulation capacity of the tank (median) obtained taking into account uncertainty was lower than that determined by the probabilistic solution. Concurrently, the increase in the unit outflow (q) was accompanied by an increase in the absolute difference in tank capacities (median) provided by the deterministic and probabilistic solutions. Therefore, the tank capacity (median value) determined via the simulation with the hydrodynamic model, where uncertainty is excluded is underestimated in comparison with the deterministic solution. From the point of view of the operational reliability of the tank, this is a favourable result, as the tank will not be overflowed. However, given the variability of the established 95% confidence interval and the fact that the analyses were carried out for a single catchment, it is difficult to formulate far-reaching guidelines and generalisations. Hence, further investigations of urban catchments with different physical and geographical characteristics are advisable.

7.3.1.16 *Graphical-analytical-probabilistic method for rectangular tank dimensioning*

The studies mentioned above (Hong 2008; Kotowski et al. 2016; Szeląg et al. 2019) indicate that the methods for determination of the capacity of tanks with a constant cross-section of the accumulation chamber (A = const) are highly simplified and ignore a number of factors influencing the filling process. Additionally, these models assume a simplified description of the inflow hydrograph shape (triangular, rectangular, exponential functions), ignore the variability of the parameters of the inflow hydrograph shape, and do not take the period between rainfall events into account. Therefore, the sizes of the accumulation chamber and the designed drain devices may be over- or underestimated, which indicates the need to modify the methodology for determination of tank capacity. Taking the above into account, the paper proposed an original method for determination of tank capacity. Three calculation modules were distinguished in the innovative approach:

a determination of the required tank capacity for the assumed surface area of the accumulation chamber (A) and drain devices (D),
b simulation of annual and multiyear time series of tank inflow hydrographs taking into account their different shapes,
c control of tank overfilling in terms of the time of tank emptying and the occurrence of several intense rainfall events at short time intervals.

The calculation algorithm of the proposed model is presented in Figure 7.26. It indicates that the simulation of the stochastic nature of the inflow hydrograph variability should be based on continuous time series $Q(t)$. They are used for determination of hydrograph parameters, which so far has been possible to a limited extent in probabilistic models. The proposed model for tank capacity prediction can be divided into five independent components. These include a simulator of tank capacity as well as the filling and emptying time, a module for prediction of the reliability of tank operation in the assumed calculation period, a module for simulation of hydrograph

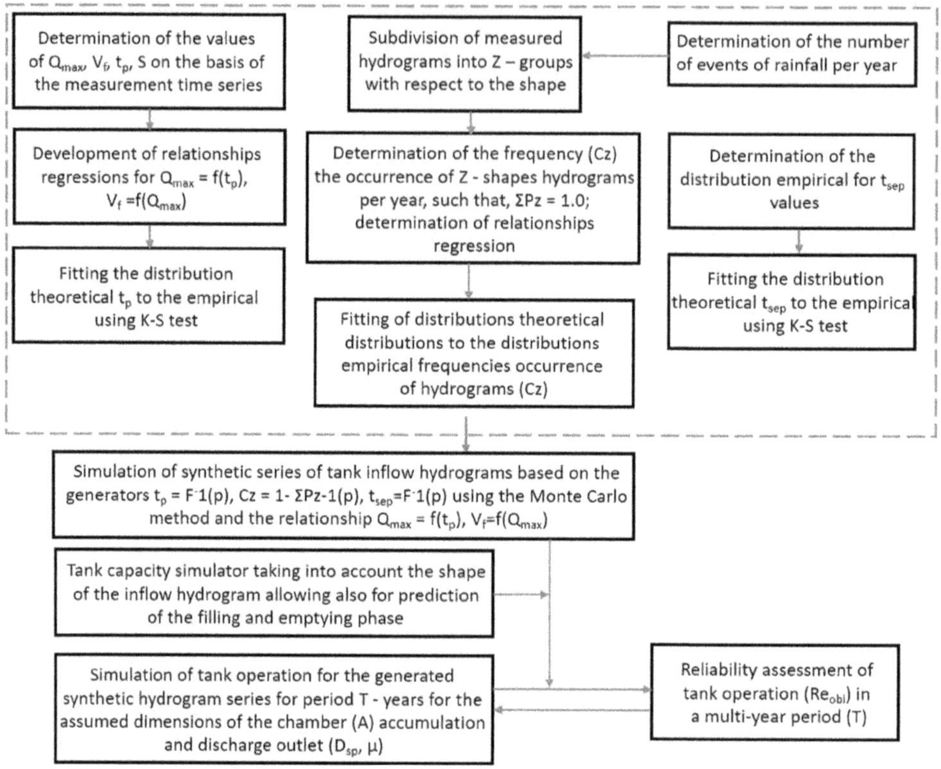

Figure 7.26 Algorithm of the model for tank capacity prediction.

time series, and a module for determination of the probabilistic characteristics of distributions describing the variability of inflow hydrograph parameters.

The simulator of the capacity of the retention tank is an analytical-graphical model used for determination of the desired tank operation parameters without the need to solve a system of differential equations. The computational module for assessment of reliability helps to determine the number of tank overfilling events in the assumed period, which may be one of the criteria for selection of the geometry of the accumulation chamber and the dimensions of the drain devices. One of the most complex components of the model is the module for characterisation of probabilistic distributions describing the variability of the inflow hydrograph parameters. In this module, based on hydrograph measurement results, the basic shape parameters are determined (peak flow, volume, rise time, unevenness of the stormwater depth distribution), regression models are constructed, empirical distributions are determined, and theoretical distributions are selected in order to simulate the probabilistic nature of hydrograph variability.

Another module facilitates modelling the time series of tank inflow hydrographs based on determined regression relationships and the hydrograph parameter generators developed with the Monte Carlo method.

In the case of a single-chamber tank with a constant cross-section of the accumulation chamber (A = const) where the outflow proceeds through a circular drain, the differential equation of the volume balance can be expressed in the following general form:

$$\frac{dV}{dt} = Q_{\max} \cdot f(t) - C_z \cdot V^{0.50} \tag{7.59}$$

where: $Q_{\max}$ – maximum stormwater inflow to the tank, C_z – parameter taking into account the cross-section of the accumulation chamber and the drain devices determined with the formula:

$$C_z = \frac{\mu \cdot \pi \cdot D_{sp}^2 \cdot \sqrt{2 \cdot g}}{4 \cdot \sqrt{A}} \tag{7.60}$$

where: μ – discharge coefficient, D_{sp} – diameter of the drain orifice, A – surface area of the accumulation chamber, $f(t)$ – tank inflow function described by the following relationships:

- triangular hydrograph:

$$\text{for } t \le t_p \quad Q(t) = \frac{Q_{\max}}{t_p} \cdot t \tag{7.61}$$

$$\text{for } t \ge t_p \quad Q(t) = Q_{\max} - \frac{Q_{\max}}{\lambda \cdot t_p} \cdot (t - t_p) \tag{7.62}$$

- unevenness of the stormwater volume distribution in the hydrograph:

$$S = \frac{1}{\lambda + 1} \tag{7.63}$$

- hydrograph expressed as a power function

$$\text{for } t \le t_p \quad Q(t) = Q_{\max} \cdot \left(\frac{t}{t_p}\right)^{a_1} \tag{7.64}$$

$$\text{for } t \ge t_p \quad Q(t) = Q_{\max} \cdot \left(\frac{\lambda \cdot t_p - t}{\lambda \cdot t_p}\right)^{a_2} \tag{7.65}$$

- unevenness of the stormwater depth distribution in the hydrograph:

$$S = \frac{\dfrac{1}{a_1 + 1}}{\dfrac{1}{a_1 + 1} + \dfrac{\lambda}{a_2 + 1}\left(\left(\dfrac{1}{\lambda}\right)^{a_2 + 1} - \left(\dfrac{\lambda - 1}{\lambda}\right)^{a_2 + 1}\right)} \tag{7.66}$$

- hydrograph expressed as the gamma distribution:

$$Q = Q_{\max} \cdot \left\{\left(\frac{t}{t_p}\right) \cdot \exp\left(1 - \frac{t}{t_p}\right)\right\}^{n-1} \tag{7.67}$$

where: a_1, a_2, λ, n – parameters of the inflow hydrograph shape, t_p – time to reach the maximum flow.

In order to transform Eq. 7.59 into the dimensionless form, the following variables are introduced:

- normalised time from the beginning (t^*):

$$t^* = \frac{t}{t_p} \tag{7.68}$$

- normalised inflow to the tank (Q^*)

$$Q^* = \frac{Q(t)}{Q_{max}} \tag{7.69}$$

- normalised tank capacity (V^*):

$$V^* = \frac{V(t)}{V_c} \tag{7.70}$$

where: V_c – total volume of the inflow hydrograph.

By substitution of Eq. 7.68–7.70 into Eq. 7.59 after appropriate transformations, the following equation is obtained:

$$\frac{dV^*}{dt^*} = \omega \cdot f\left(t^*\right) - R \cdot \left(V^*\right)^{0.50} \tag{7.71}$$

where: ω – shape parameter defined as $Q_{max} \cdot t_p \cdot V_c^{-1}$, R – parameter linking the tank and inflow hydrograph characteristics determined with the formula:

$$R = \frac{\mu \cdot \pi \cdot t_p \cdot D_{sp}^2 \sqrt{2 \cdot g}}{4 \cdot \sqrt{A \cdot V_c}} \tag{7.72}$$

Assuming different shapes of inflow hydrographs and the relationship $R = f(\omega, \beta)$, the differential Eq. 7.71 can be solved. Figure 7.27 shows an example of the relationship $R = f(\omega, \beta)$ for a triangular inflow hydrograph.

The curves in Figure 7.27 indicate that the value of parameter R is a linear function of ω for individual values of β. Moreover, to maintain a constant value of R while ω is increasing ($\omega_2 > \omega_1$), it is necessary to reduce the maximum flow reduction coefficient, which in practice involves reduction of the drain outlet diameter (D_{sp}) and/or the discharge coefficient (μ) assuming a constant surface area (A).

When the assumed value β is different than in graph $R = f(\omega, \beta = \beta^*, \beta^* = \text{const})$, it is necessary to interpolate the value, which may produce an error. To avoid the error, relations $R = f(\omega)$ for the individual values of the maximum flow reduction coefficients (β) were approximated with the $R = \theta(\beta) \cdot \omega$ function. The calculations indicated that approximation of the variability $R = f(\omega, \beta = \beta^*, \beta^* = \text{const})$ with a linear function was accompanied by changes in the values of the coefficients of determination (R^2) for $\beta = 0.1 \div 0.9$ in the range of $0.996 \div 0.999$.

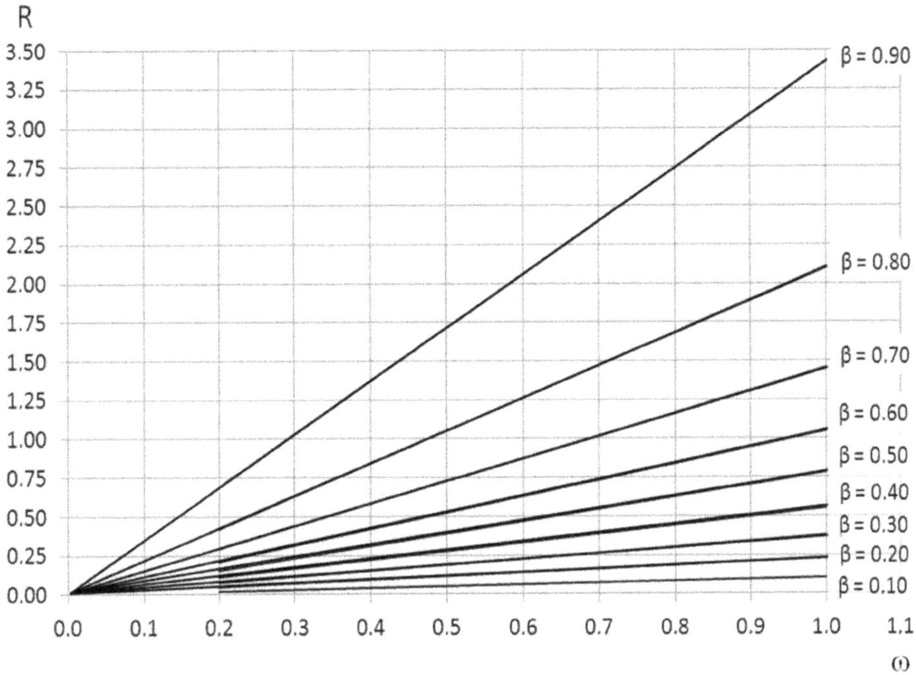

Figure 7.27 Impact of the hydrograph asymmetry factor (ω) and the maximum flow reduction coefficient (β) on the value of parameter (*R*) for a triangular inflow hydrograph (Based on Szeląg & Mrowiec 2016).

This confirmed the high agreement of the results provided by the generated nomograms with the results of numerical calculations. Hence, the values of parameters θ were determined for the analysed inflow hydrographs, and a graph showing the impact of β on the value of θ was prepared (Figure 7.28). On the basis of the relation $\theta = f(\beta)$, it is possible to determine the value of parameter θ necessary for calculation of the value of the function $R = f(\omega)$, which is the basis for calculation of the maximum flow reduction coefficient.

The graph in Figure 7.29, prepared taking into account the SWMM simulation results, shows the impact of the inflow hydrograph asymmetry factor and the maximum flow reduction coefficient on the dimensionless retention tank capacity (V_{zb}/V_c). It was indicated by the graph that the designed tank capacity reached its maximum value when the share of the rise phase depth in the total hydrograph volume (triangular, described by the gamma distribution or power function) was $S = 0.5$. Moreover, it was shown for the considered hydrograph shapes that an increase in the asymmetry factor S leads to an increase in the required capacity of a single-chamber tank. When the shape of the catchment inflow hydrograph can be described by the gamma distribution, the change in the unevenness of the stormwater depth distribution (S) has the greatest impact on the tank capacity (Figure 7.29). In turn, an insignificant effect of S on V_{zb}/V_c was observed for the variability $Q_d = f(t)$ described by the triangle. For example, for $\beta = 0.4$, the change in the value of S from 0.1 to 0.5 for the hydrograph described by the gamma distribution and by the triangle contributes to an increase in V_{zb}/V_c from 0.40 to 0.57 (42.5%) and from 0.47 to 0.53 (13%), respectively. The smallest tank volume among the analysed hydrograph shapes is achieved in the case of a rainfall intensity distribution determining the occurrence of

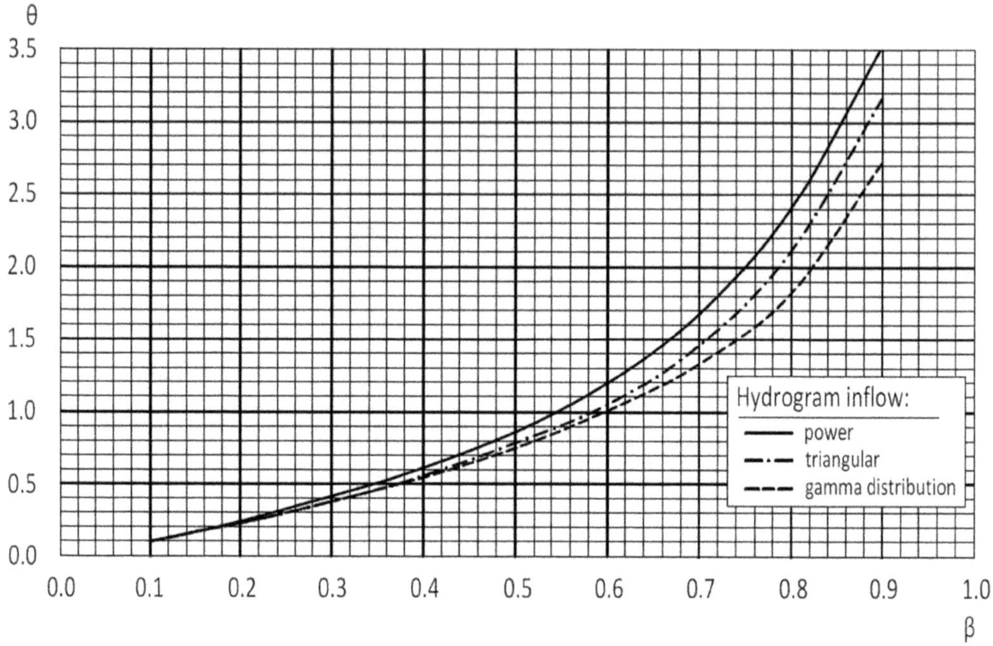

Figure 7.28 Dependence of parameter $\theta = f(\beta)$ (Based on Szeląg & Mrowiec 2016).

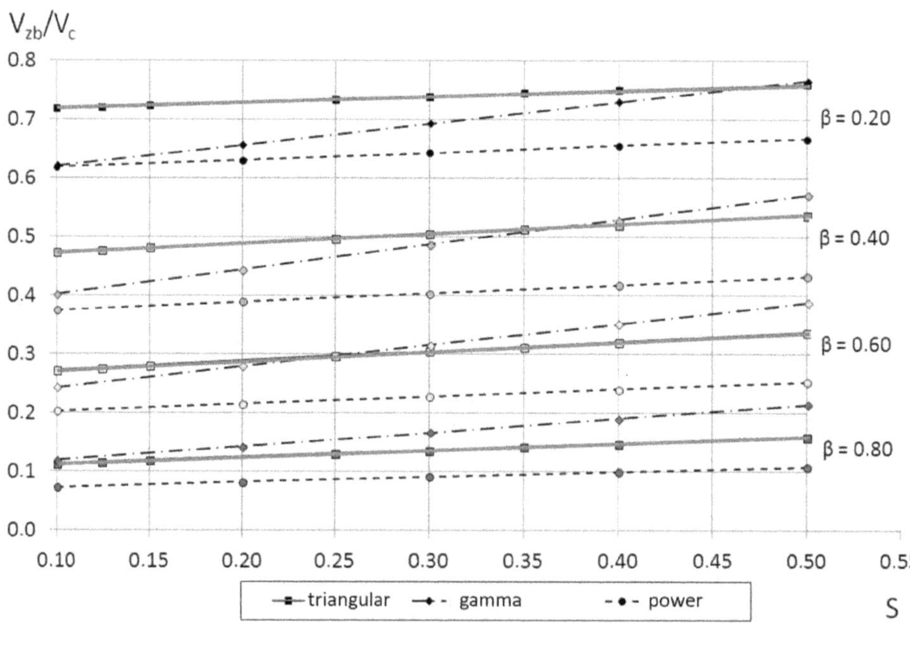

Figure 7.29 Effect of the hydrograph asymmetry factor (S) and the maximum flow reduction coefficient (β) on the dimensionless retention tank capacity (V_{zb}/V_c) for a triangular inflow hydrograph and a hydrograph described by the power function and gamma distribution (Based on Szeląg & Mrowiec 2016).

a catchment outflow hydrograph defined by the power function. When the variability $Q_d = f(t)$ can be expressed by the gamma distribution, the calculated accumulation volumes for $\beta = 0.2$, $\beta = 0.4$, $\beta = 0.6$, and $\beta = 0.8$ are larger than V_{zb} obtained for the triangular hydrograph at $S>0.47$, $S>0.37$, $S>0.26$, and $S>0.10$. Moreover, when the inflow to the tank can be described by the triangular hydrograph, the largest required accumulation volume with respect to the shapes considered in the study is achieved for $\beta<0.2$ in the whole range of values of the analysed asymmetry factor S. When the variability $Q_d = f(t)$ can be expressed by the gamma distribution, the dimensionless tank volume for $\beta = 0.8$ is larger than the V_{zb}/V_c value determined for a triangular hydrograph and a hydrograph described by the gamma distribution (Figure 7.29).

Besides the tank capacity, significant practical importance is ascribed to the hydraulic characteristics describing the tank function, i.e., the filling and emptying time. These aspects are relevant at the stage of tank design, as it is essential to prevent tank overflowing within a short time in the case of two consecutive rainfall events. Assuming that the tank outlet is not flooded, the emptying time can be determined with the following equation:

$$t_{\text{opr}} = \frac{2 \cdot V_{\text{zb}}}{4^{-1} \cdot \mu \cdot \pi \cdot D_{\text{sp}}^2 \sqrt{\frac{2 \cdot g}{A}}} = \frac{2 \cdot V_{\text{zb}}}{C_z} \tag{7.73}$$

In turn, assuming that the time between two rainfall events is equal to t_{sep} and the tank will not empty completely after this period, the volume of the remaining stormwater can be determined as follows:

$$V_0 = \left(V_{\text{zb}}^{0.5} - 4^{-1} \cdot \mu \cdot \pi \cdot D_{\text{sp}}^2 \sqrt{\frac{2 \cdot g}{A}} \cdot t_{\text{sep}} \right)^2 \tag{7.74}$$

The filling time is another key aspect in proper tank functioning. Knowing that the maximum accumulation capacity is reached after the occurrence of the maximum inflow to the tank, during the decrease in the inflow intensity, the filling time can be determined with the following equation:

$$Q(t) = Q_{o,\max} = \mu \cdot \frac{\pi \cdot D_{\text{sp}}^2}{4} \cdot \sqrt{2 \cdot g \cdot \frac{V_{\text{zb}}}{A}} \tag{7.75}$$

On the basis of this formula, the filling time in the case of an inflow to the tank can be expressed as:

- triangular inflow hydrograph:

$$t_{\text{nap}} = t_p + \lambda \cdot t_p \cdot \frac{Q_{\max} - Q_{o,\max}}{Q_{\max}} \tag{7.76}$$

- power function:

$$t_{\text{nap}} = \lambda \cdot t_p - \lambda \cdot t_p \cdot \left(\frac{Q_{o,\max}}{Q_{\max}} \right)^{\frac{1}{a_2}} \tag{7.77}$$

In the case of an inflow hydrograph described by the gamma distribution function, it is not possible to formulate an analytical relationship due to the entangled form of equation $Q = f(t)$.

7.3.1.17 Reliability of tank operation

The reliability of a facility is reflected by the probability of occurrence of an event described as follows (Price & Vojinovic 2011):

$$R(t) = p(t > \tau) \tag{7.78}$$

where: $R(t)$ – reliability, t – operation time of the facility without failure, τ – required operation time without failure.

Thus, reliability is one of the important measures for assessment of the operation of sewer networks and their facilities. At the tank dimensioning stage, a number of criteria can be introduced to assess the reliability of its operation. The important criteria for evaluation of tank performance include:

- overflowing,
- number of discharges via emergency overflow,
- volume of discharges via emergency overflow.

Concurrently, with reference to the definition of reliability, a number of reliability criteria can be defined:

$$R(t)_p = \frac{N_p}{N_t} \tag{7.79}$$

$$R(t)_z = \frac{N_z}{N_t} \tag{7.80}$$

$$R(t)_V = \frac{V_z}{V_t} \tag{7.81}$$

where: N_p – number of tank overflows in the period T covered by the calculations, N_t – number of rainfall events in the period covered by the simulations, V_z –volume of discharges via the emergency overflow, V_t – total inflow hydrograph volume in the period covered by the simulations.

It was assumed in these analyses that tank overfilling occurs when the following condition is met:

$$V_0(t) + V_{zb}(Q_{max}, t_p, V_c; t+1) > V_{zb} \tag{7.82}$$

and the discharge via the emergency overflow has the following volume:

$$V_V = V_0(t) + V_{zb}(Q_{max}, t_p, V_c; t+1) - V_{zb} \tag{7.83}$$

These criteria can be applied individually and in combination to form the basis for assessment of the reliability of the retention tank operation.

Determination of probabilistic distributions of the variability of inflow hydrograph parameters.

On the basis of the analyses described above, it can be concluded that the inflow hydrograph shape has an impact on the tank capacity and operating parameters. The analyses and literature review (Hong 2008; Wang & Guo 2018) show that tank functioning is mainly influenced by the maximum instantaneous flow (Q_{max}), rise time (t_p), inflow hydrograph volume (V_f), and unevenness of the stormwater depth distribution ($S = V_p \cdot V_c^{-1}$). These parameters exhibit great variability in the measured hydrographs and have a probabilistic character. In order to simulate their variability, multivariate distributions are generated based on marginal distributions (theoretical distributions obtained from empirical ones determined by measurements). Marginal distributions are combined using special functions (Osorio et al. 2009; Vernieuwe et al. 2015); however, to obtain a reliable multidimensional distribution, a sufficiently large number of measurement data are necessary. The limitations in the number of data in such a complex model may lead to loss of their reliability, which impacts calculation results. Therefore, at the stage of model generation in practice, the analysed variables are correlated in order to describe the inflow hydrographs, and the number of probabilistic variables in the model requiring simulation is reduced. The simplifications introduced at the stage of calculations result in omission of phenomena that may have a large impact on simulation results.

In the computational methodology, it was assumed that the inflow hydrograph parameters are determined based on the following empirical relationships:

$$Q_{\max} = b_1 \cdot t_p + b_0 \tag{7.84}$$

$$V_c = c_1 \cdot Q_{\max} + c_0 \tag{7.85}$$

where: t_p – time to reach the maximum flow in the hydrograph, i.e., a probabilistic variable generated based on theoretical distributions determined with the Monte Carlo method, which can be expressed as $t_p = F^{-1}(p)$.

By substitution of (7.34) and (7.35) into the equation for the ω value in Eq. 7.71, the probabilistic parameter of the inflow hydrograph shape can be defined as follows

$$\omega = \frac{F^{-1}(p)\cdot\left(b_1 \cdot F^{-1}(p) + b_0\right)}{c_1 \cdot \left(b_1 \cdot F^{-1}(p) + b_0\right) + c_0} \tag{7.86}$$

Besides the variability of the ω value and its probabilistic character, the calculations take into account different inflow hydrograph shapes limited to three typical ones, as discussed by Szeląg (2013). Triangular hydrographs (Z_1) and hydrographs described by gamma distribution (Z_2) and by power functions (Z_3) were used in the paper. The key issue in the model is to determine empirical distributions and those based on theoretical values of $Z_1 = F_{Z1}^{-1}(p)$, $Z_2 = F_{Z2}^{-1}(p)$, and $Z_3 = F_{Z3}^{-1}(p)$ expressing the probability a given shape of the inflow hydrograph. To obtain these relationships, it is important to determine the mean number of rainfall events per year (K), which allows defining the series of $F_{Zi}^{-1}(p)$ values in such a way that the following conditions are met:

$$F_{Z1}^{-1}(p) + F_{Z2}^{-1}(p) + F_{Z3}^{-1}(p) = 1 \tag{7.87}$$

$$K \cdot F_{Z1}^{-1}(p) + K \cdot F_{Z2}^{-1}(p) + K \cdot F_{Z3}^{-1}(p) = K \tag{7.88}$$

In this way, the generated inflow hydrographs forming the time series of rainfall events K are described by the following parameter:

$$\omega\left(F_{Z1}^{-1}(p)\right)_{p=1,2,\ldots,K} = \frac{F^{-1}(p)\cdot\left(b_1 \cdot F^{-1}(p) + b_0\right)}{c_1 \cdot \left(b_1 \cdot F^{-1}(p) + b_0\right) + c_0} \tag{7.89}$$

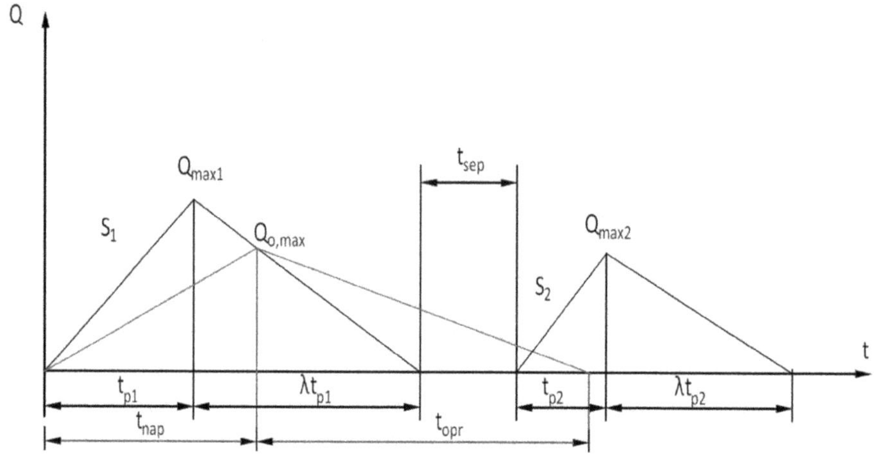

Figure 7.30 Calculation scheme for retention tank overfilling.

In a simplified case, the construction of the model describing the time series of inflow hydrographs based on parameter (7.89) can be completed, which is usually done in probabilistic models for simulation of the tank capacity. Nevertheless, this approach ignores the time between successive rainfall events. In the case of two intense rainfall events occurring within a short time, this may result in tank overfilling, which is related to the tank emptying time (Figure 7.30).

To take into account the potential occurrence of such events in simulations of the retention tank operation, it is necessary to include the separation time (t_{sep}) between successive hydrographs and another stochastic variable $t_{sep} = F_s^{-1}(p)$. Thus, successive K values of the events in the hydrograph time series generate the following sequences:

$$\omega\left(F_{Z1}^{-1}(p);F_s^{-1}(p)\right)_{p=1,2,\ldots,K} = \frac{F^{-1}(p)\cdot\left(b_1\cdot F^{-1}(p)+b_0\right)}{c_1\cdot\left(b_1\cdot F^{-1}(p)+b_0\right)+c_0}\left(F_{Z1}^{-1}(p);F_s^{-1}(p)\right) \tag{7.90}$$

Substitution of formulas (7.26)–(7.32) to (7.14) and inclusion of $R = \theta(\beta)\cdot\omega$ give a stochastic equation for determination of the retention tank capacity:

$$Q_{o,\max}\cdot\frac{t_p}{\sqrt{V_c}}\left(\omega\left(F_{Z1}^{-1}(p);F_s^{-1}(p)\right)_{p=1,2,\ldots,K}\right)=$$

$$=Q_{o,\max}\cdot\frac{F^{-1}(p)}{\sqrt{c_1\cdot\left(b_1\cdot F^{-1}(p)+b_0\right)+c_0}}\left(\omega\left(F_{Z1}^{-1}(p);F_s^{-1}(p)\right)_{p=1,2,\ldots,K}\right)$$

$$=\theta\left(\frac{Q_{o,\max}}{\left(b_1\cdot F^{-1}(p)+b_0\right)}\right)\omega\left(F_{Z1}^{-1}(p);F_s^{-1}(p)\right)_{p=1,2,\ldots,K}$$

$$Q_{o,\max} \cdot \frac{t_p}{\sqrt{V_c}} \left(\omega\left(F_{Z1}^{-1}(p); F_s^{-1}(p)\right)_{p=1,2,\ldots K} \right)$$
$$= \theta \left(\frac{Q_{o,\max}}{\left(b_1 \cdot F^{-1}(p) + b_0\right)} \right) \frac{F^{-1}(p) \cdot \left(b_1 \cdot F^{-1}(p) + b_0\right)}{c_1 \cdot \left(b_1 \cdot F^{-1}(p) + b_0\right) + c_0} \left(F_{Z1}^{-1}(p); F_s^{-1}(p)\right) \quad (7.91)$$

On the basis of Eq. 7.91, it is possible to determine the $Q_{o,\max}$ value for the assumed geometry of the accumulation chamber (A), drain devices (D_{sp}), and the stochastic inflow hydrograph. The value can be used for determination of tank filling. Moreover, it is possible to determine whether the designed tank will overflow during the period covered by the calculations, which is important for the operation of the tank and regulation of the volume of the stormwater inflow stream $Q(t)$.

7.3.1.18 Numerical methods

Numerical methods for determination of tank capacity are based on appropriate calculation schemes, which can be divided into the so-called explicit and implicit methods. Since the Saint Venant system of differential equations (Eqs. 7.30 and 7.31) is used in the case of open-field reservoirs, velocity distributions in the cross-section are determined in addition to the dynamics of filling. For this purpose, it is necessary to use complex numerical algorithms described by, e.g., Szymkiewicz (2010). In the case of tanks located in stormwater sewer systems, the initial stormwater depth is usually ignored; hence, their capacity is calculated with the differential equation of the stormwater volume balance Eq. 7.4, which is solved with the use of the so-called explicit and implicit methods. In the former methods, the value of function $y_{n+1} = F(x_n, y_n, y_n')$ at time point $t+1$ is determined from the value at time t at point x. In the latter approach, the value of $y_{n+1} = G(x_n, y_{n+1}, y_{n+1}')$ is determined iteratively. In the design of retention tanks as part of sewer networks, the differential equation of the stormwater volume balance (Eq. 7.4) is solved with the finite difference methods proposed by Euler, 4th order Runge-Kutta method, or Adam-Bashforth (Szymkiewicz 2000).

Over the last 30 years, the considerable progress in computational methods has largely simplified the solution of complex systems of differential equations. This has led to the development of programs for modelling surface runoff in both urban and agricultural catchments. Well-known programs includes SWMM (Storm Water Management Model), MOUSE, MIKE URBAN, SIMBA, etc. Additionally, given the important role of tanks in the functioning of sewer networks, the programs provide appropriate toolboxes for the determination of retention tanks. In the simplest case, the SWMM tank model consists of four components, i.e., an inlet channel, an accumulation chamber, a drain, and an outflow (Figure 7.31). Figures 7.32 and 7.33 show individual dialog boxes of the Storm Water Management Model program, in which the necessary characteristics required for a tubular tank design are defined. First, the reservoir inflow and outflow nodes should be entered (Figure 7.32), simultaneously providing the bottom ordinates, the maximum depth of the stormwater in the nodes, and the ordinate of the outlet from the analysed sewer network section.

In the next step, the accumulation chamber of the tank, i.e., the diameter, length, and roughness coefficient, and dimensions of the orifice with the corresponding flow coefficient should be defined.

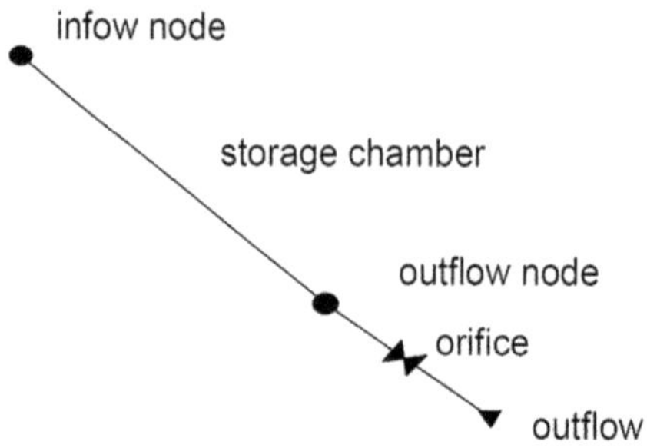

Figure 7.31 Tank calculation scheme in the SWMM program.

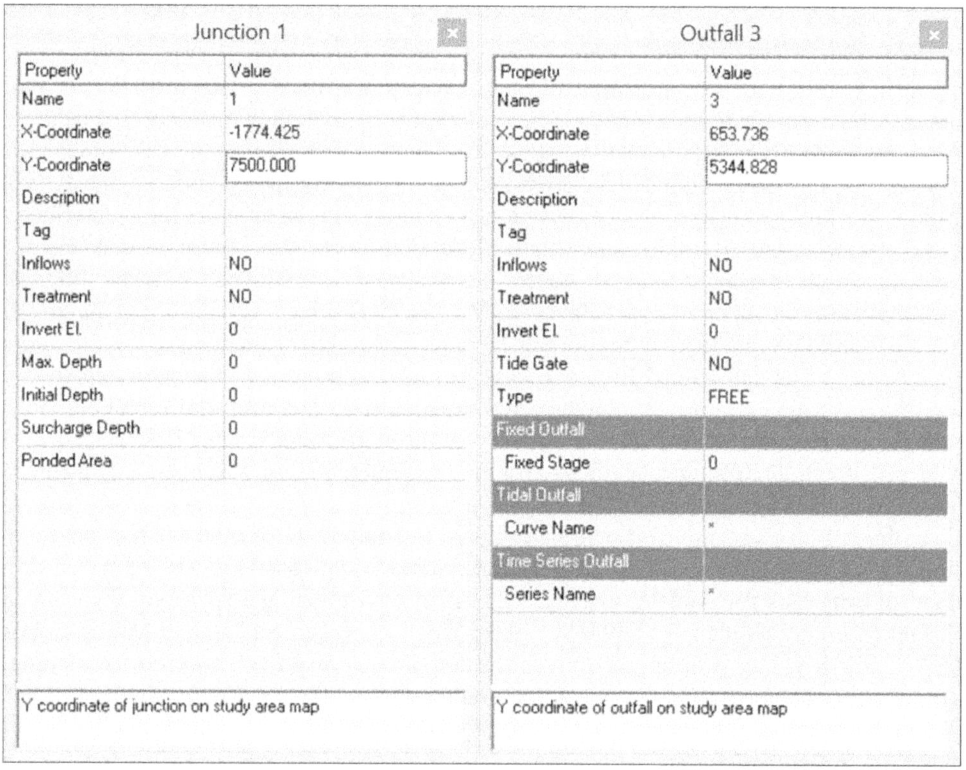

Figure 7.32 Dialog boxes in the SWMM program for (a) inflow/outflow nodes, (b) sewer network outlet, (c) accumulation chamber, (d) diameter of the drain hole.

When the retention of the inflow channel is omitted, the hydrograph can be defined as the node inflow. In such a model, the calculation step is determined and tank filling simulations are performed until the accumulation chamber filling is equal to the diameter of the designed tubular tank.

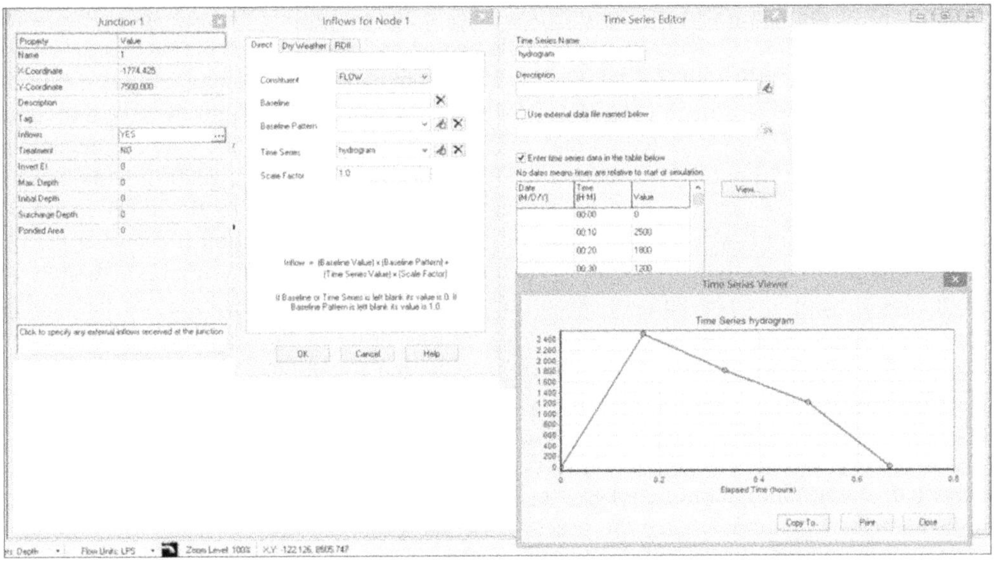

Figure 7.33 Dialog boxes in the SWMM program for the node inflow hydrograph.

In computer programs, the retention tank can be integrated with the hydrodynamic catchment model, which gives an opportunity to study the impact of the rainfall-runoff phenomenon on its operation and to analyse the impact of the so-called uncertainty of hydrodynamic model parameters on the functioning of facilities located in the sewer network.

When stormwater overflows onto the ground surface and the time of stormwater flow through the network is longer than 15 minutes, the calculation of the required tank accumulation capacity is based on long-term simulations in hydrodynamic models (DWA-A 117 2006) with the use of long-term rainfall observation series (minimum 30 years). In this approach, known as the generalised method, the calculations aim to determine the frequency of surcharge of the capacity of analysed tanks. When the assumed frequencies do not meet the relevant requirements, correction of the capacity and repeated simulation are advisable. The time series used for calculations should not be discontinuous or should be completed with appropriate methods in the case of discontinuity (Stalmann 1991; Verworn 1992; Einfalt & Wolf-Schumann 1992). Moreover, in the hydrodynamic model, the DWA-A 117 guidelines impose the necessity to recreate the runoff function for both impervious and pervious areas, and it is recommended to include the phenomenon of stormwater infiltration into the ground. The simulation results obtained in this way can be the basis for determination of the relationship between the required accumulation capacity and the probability of surcharge.

The presented analysis allows the formulation of a conclusion that the methods described can be used to calculate the capacity of a tubular tank using only a triangular and trapezoidal inflow hydrograph as the basis for calculations. The schematisation of the tank outflow hydrograph by means of a triangle or a trapezoid may result in underestimation of the accumulation chamber capacity and tank overfilling, as confirmed by the simulation results. An interesting development of the methods for tank dimensioning is the simplified method specified by the DWA-A 117 guidelines, in which the tank design stage includes a risk function increasing the tank capacity (f_z) and reducing (f_A), taking into account phenomena accompanying surface run-off and stormwater flow through channels. Exclusion of the rainfall dynamics during individual rainfall events in the graphical method may lead to underestimation of the required accumulation capacity, as confirmed by calculation results reported by Mrowiec (2009). Although the author introduced modifications to the graphical method, rainfall observational series are the basis for its application, which in many cases is a considerable limitation preventing the use of the method. Since analytical-graphical methods are based on a differential equation transformed into finite differences and are part of the methodology of calculation of the tank operation, the results obtained with these methods are similar to those from the numerical solution. However, the maximum accumulation capacity is recorded at complete chamber filling and the $V = f(h)$ curve depends on the length of the retention chamber. Therefore, the calculations in the analytical-graphical method should be performed with the method of successive approximations (assuming different lengths of the tank) until the determined tank filling is equal to the diameter of the accumulation chamber. Among the tank dimensioning approaches described above, numerical methods ensure the greatest accuracy, as they are based on the solution of the differential equation of the volume balance with the use of relevant numerical schemes. Although the analytical-probabilistic methods take into account the stochastic nature of rainfalls, they can be used in small urban catchments due to the schematisation of the inflow and outflow hydrographs (rectangle or triangle). In these methods, the assumption of the absence of correlations between rainfall parameters (rainfall depth, duration, dry period) is not justified when there is a statistically significant relationship between these rain characteristics (Balistrocchi & Bacchi 2011). In this case, it is advisable to use copula functions for development of a multidimensional distribution of probability density, which is the basis for simulation of rainfall series in the analysed period. Additionally, given the stochastic character of the catchment and sewer system parameters, it is advisable to investigate the effect of the parameters of hydrodynamic models on the tank accumulation capacity.

7.3.1.19 Flow regulators

The application of hydrodynamic modelling of catchments and stormwater networks provides the possibility of implementation, in the tasks of controlling the volume of run-off from reservoirs, of flow regulators in drainage systems (provided that the hydraulic conditions in the catchment above allow for delaying the run-off from the catchment and reducing the flow) in real time. The real time control (RTC) system is aimed at maximising the use of existing infrastructure and reducing capital costs. The greatest advantage of RTC systems is the ability to control the operation of key elements of the stormwater network by adjusting the system to the hydraulic load resulting from the temporal and spatial variability of rainfall in the catchment. Since the control of retention capacity constitutes the basis of RTC, these systems are associated with retention reservoirs and the ability to regulate outflow from them in a dynamic system. Classification of RTC systems can be made on the basis of two criteria:

- reactive, where reservoir/regulator operating conditions and reservoir outflows are determined based on the current hydraulic load,

- simulation-based, where rainfall data are modelled in advance using simulation-based machine learning methods, which provides an opportunity to determine the reservoir outflow conditions in advance.

In the case of extensive stormwater networks where performance problems are local, it is not necessary to implement a catchment-wide RTC system; thus, three types of solutions are distinguished:

- local, in which the control concerns a single or a few devices influencing the hydraulic conditions in a part of the stormwater network,
- global, in which the data from measuring and monitoring devices are collected and constitute the input data for computer simulations as a basis for decision making,
- integrated, in which monitoring and control cover simultaneously the stormwater network, wastewater treatment plant, and the receiver.

Depending on the requirements given the objectives of RTC systems, they can be classified as:

- quantitative, aimed at reducing the incidence of flooding,
- qualitative, optimising the load entering the wastewater treatment plant and reducing energy consumption, greenhouse gas emissions, etc,
- immission-related, in which the operating parameters of the stormwater network and wastewater treatment plants are determined on the basis of measurements of the receiving water quality.

Selection of the RTC system and the need for its implementation are determined individually for each catchment, depending on the requirements for such a system. Therefore, the guidelines developed for the evaluation and need for implementation of the RTC system take into account: catchment size, land use, underground infrastructure (overflows, retention basins, flow regulators, pumping stations), channel slopes, number of outlets to the receiving water body, impact of the stormwater system on the receiving water body, receiving water quality and requirements, size of the treatment plant and the applied wastewater treatment technology.

Basin outflow control can be represented as a control loop including: process monitoring sensors, controllers that affect the dynamics of the considered process, controllers and a data communication system. In this system, the control loops are implemented in the following way:

- feedback, the controller settings depend on a certain deviation compared to the set value,
- anticipatory feedback, the controller sets the values in advance, which is determined by the stormwater system simulation tool used,
- a combination of both.

The available computational tools, including SWMM, allow implementing controller characteristics. If there are numerous reservoirs in a stormwater network, real-time outflow control is a very complex task, requiring optimisation of the settings. Real time control requires very fast simulation tools; thus, a compromise is sought between a reliable representation of the prevailing conditions using hydrodynamic models and simulation time. Simplifications are also sought

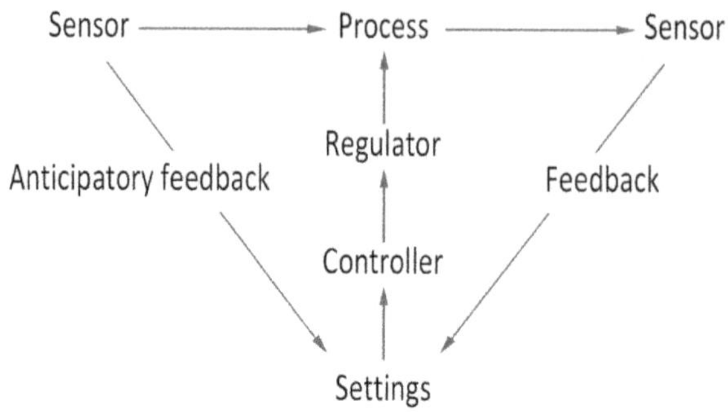

Figure 7.34 Operation of the stormwater system and process optimisation using RTC.

in the models describing catchment run-off, hydraulic conditions in the stormwater network, and interactions in the rainfall – run-off – stormwater flow – network overloading/stormwater flooding. Due to the fact that the execution time of a single simulation of stormwater system operation takes several seconds in the RTC system, this is unacceptable and alternatives to hydrodynamic models are needed. Observing the development of computational techniques, a great potential can be seen in machine learning (Szeląg et al. 2020). In the RTC system, on the basis of the so-called soft-sensor developed, it is possible to model the operation of the stormwater system and optimise the process in tenths and hundredths of a second (Figure 7.34).

Chapter 8

Case studies

8.1 Sensitivity analysis and global sensitivity analysis in a selected urban catchment model

8.1.1 Object of analysis

This chapter presents an example of calculations in sensitivity analysis and probabilistic calibration of an urban catchment model located in the city of Kielce (Figure 8.1). The catchment, with a total surface area of 62 ha, is located in the central-eastern part of the city.

The main channel is 1,569 m long and its diameter in the section from the inlet to the outlet varies from 600 to 1,250 mm. The channel receives stormwater from 17 side channels, and the total length of the network is 5,583 m. The sewer system depth varies in the range of 1.59–3.55 m. The catchment area comprises mainly housing estates, public utility buildings, and main and side streets. In total, roofs account for 14.3%, pavements for 8.4%, and parking lots for 11.2% of the catchment area. The surface area of roofs with direct drainage to the channels constitutes 72.5% of the entire roof area. The approximate capacity of the channels (without manholes and inlets) in the collector system is 2,032 m^3.

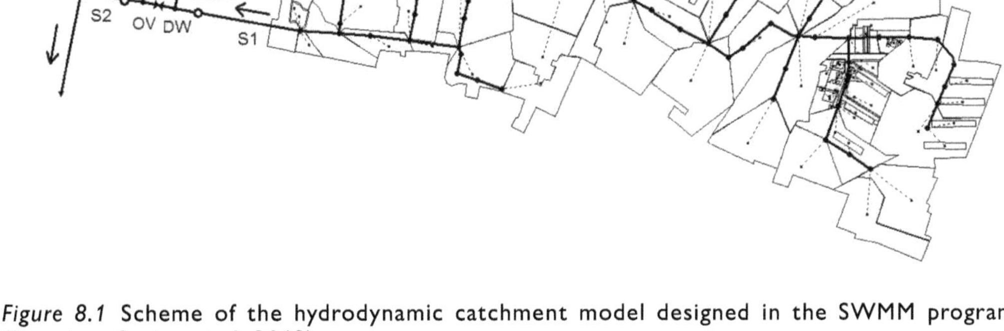

Figure 8.1 Scheme of the hydrodynamic catchment model designed in the SWMM program (Based on Szeląg et al. 2019).

DOI: 10.1201/9781003546436-8

Stormwater from the catchment area is discharged through the S1 channel to the diversion chamber (DW), and some part is discharged directly to the stormwater treatment plant (STP) until the filling height of $h_m = 0.42$ m. After exceeding the h_m value, the stormwater is discharged via the stormwater overflow (OV) into channel S2, which discharges the stormwater into the Silnica River.

An ultrasonic flow meter MES1 is installed in the channel at a distance of 3.0 m from the outlet of channel S1 to the diversion chamber (DW). In 2016, the MES 2 flow meter was installed in the outlet channel discharging stormwater to the Silnica River. Similar to MES 1, the meter conducts measurements at a 30-second step during heavy rainfall. As part of continuous monitoring, rainfall volumes are measured with the MES2 rain gauge located approximately 2 km from the northern border of the catchment.

8.1.2 Hydrodynamic model catchment

The hydrodynamic catchment model (Figure 8.1) developed in the SWMM 5.0.022 (Storm Water Management Model) program was used to calculate the stormwater depth. A comprehensive description of the tool used can be found in the study by Huber, Dickinson, and Rossmann (Huber & Dickinson 1988; Rossmann 2004). The model was previously subjected to deterministic calibration (without taking into account the uncertainty of parameters); its detailed parameters can be found in the study conducted by Szeląg et al. (2013). The model consists of 92 sub-catchments with areas ranging from 0.12 to 2.10 ha, 200 manholes, and 72 channel sections. The proportion of impervious area in the individual subcatchments ranges from 5% to 95%, and accounts for 53% of the total catchment area. The results of rainfall and flow measurements from 2009 to 2011 were used for calibration and validation (probabilistic GLUE+GSA method).

8.1.3 Results of sensitivity analysis (GSA) calculations

After calibration of the model, an attempt was made to determine the impact of the roughness coefficient (n_{imp}, n_{perv}), retention of impervious and pervious areas (d_{imp}, d_{perv}), flow path width (W), proportion of impervious area (%Imp), subcatchment slope (J), and sewer channel roughness coefficient (n_{sewer}) on the catchment inflow hydrograph. The ranges of the variability of parameters in the SWMM hydrodynamic model were determined using data reported by Fang & Ball (2007), Huber & Dickinson (1988), and Zawilski (2010). The following values were assumed in these considerations: $d_{\mathrm{imp}} = 1.0 \div 2.0$ mm, $d_{\mathrm{perv}} = 2.0 \div 5.0$ mm, $n_{\mathrm{imp}} = 0.010 \div 0.035$ $\mathrm{m}^{1/3}\cdot\mathrm{s}$, $n_{\mathrm{perv}} = 0.025 \div 0.90$ $\mathrm{m}^{1/3}\cdot\mathrm{s}$, $n_{\mathrm{sewer}} = 0.013 \div 0.03$ $\mathrm{m}^{-1/3}\cdot\mathrm{s}$, $\%\mathrm{Imp}_s = \alpha\cdot\%\mathrm{Imp}_m$, and $J_s = \beta\cdot J_m$ (where $\%\mathrm{Imp}_m$, S_m, $\%\mathrm{Imp}_s$, J_s – proportion of impervious areas and subcatchment slope determined from topographic maps and by probabilistic calibration; α, β – numerical values of parameters equal to 0.8 to 1.2 determined by probabilistic calibration). The literature describes a number of methods for determination of the flow path width. However, in many cases, they require determination of the value in each catchment individually, depending on its shape, channel arrangement in the plan, etc. In the uncertainty analysis, this poses considerable difficulty and extends the time of calculations substantially. Therefore, in these analyses, the flow path width was described as $W = \varepsilon\cdot A^{0.5}$ (where: A – subcatchment area) at $\varepsilon = 1.0 \div 2.0$. This approach facilitates identification of the main sources of uncertainty and a potential reduction of the dimensionality of the task, which is important for the effectiveness of the method.

To assess the impact of the rainfall distribution or the direction and velocity of the front movement on the inflow hydrograph, it is necessary to identify parameters of the model.

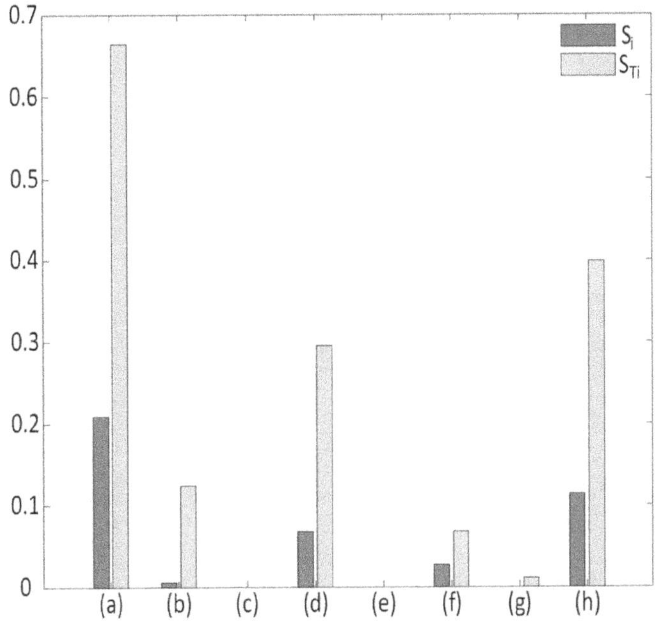

Figure 8.2 Values of sensitivity indices (S_i, S_{Ti}) for the analysed parameters (a) ε; (b) n_{imp}; (c) n_{perv}; (d) d_{imp}; (e) d_{perv}; (f) α; (g) β; (h) n_{sewer} (Based on Szeląg et al. 2016).

In the first stage of calculations, the sensitivity analysis was performed for the constructed hydrodynamic model to find parameters that have a significant impact on the catchment inflow hydrograph. This issue is important for calibration of the model, as it contributes to the limitation of the number of parameters that should be precisely identified to construct the hydrodynamic model. For the assumed ranges of numerical values of the model parameters ($n_{imp,perv}$, $d_{imp,perv}$, ε, α, β) describing the individual subcatchments, Monte-Carlo sampling was performed with 1,000 iteration, followed by simulations of the catchment inflow hydrograph. The results were used for calculation of the main sensitivity indices (S_i) and sensitivity indices resulting from interactions (S_{Ti}) between the analysed variables using (Eqs. 6.26 and 6.27), respectively; the simulation results are shown in Figure 8.2.

On the basis of the calculations (Figure 8.2), it can be concluded that the highest values of the sensitivity indices were obtained for the impervious area retention and the Manning's channel roughness coefficient, i.e. $S_i = 0.54$ and $S_{Ti} = 0.24$, as well as $S_i = 0.53$ and $S_{Ti} = 0.24$, respectively. Substantially lower values of these parameters were obtained in the case of the flow path width ($S_i = 0.09$ and $S_{Ti} = 0.03$) and the proportion of impervious area ($S_i = 0.12$ and $S_{Ti} = 0.10$). In turn, the Manning's roughness coefficient, pervious area retention, and subcatchment slope were found to have a negligible impact on the total variance of the model.

8.1.4 Results of GLUE analysis

Considering the results of the GSA calculations, the next step was to identify the model parameters, taking into account the retention and roughness coefficient of impervious areas, flow path width, and channel roughness coefficient. The other variables (n_{perv}, d_{perv}, β) were not included

Table 8.1 Characteristics of rainfall events and hydrograph volume

Date	*P_c [mm]*	*t_d [min]*	*q [dm³/(ha·s)]*	*V_c [m³]*
08.07.2011	8.6	60	23.89	1,733
15.09.2010	9.2	286	5.36	2,221
30.07.2010	12.5	107	19.47	1,908
08.07.2010	16.5	270	10.19	3,415

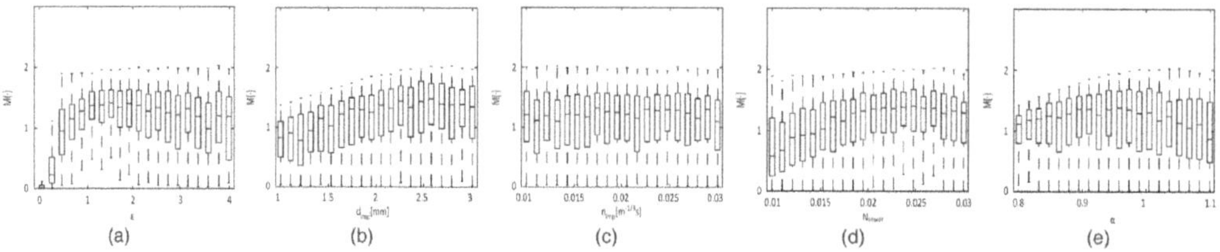

Figure 8.3 Variation of the likelihood function values for the analysed parameters (ε, d_{imp}, n_{imp}, N_{sewer}, α) in the hydrodynamic model (Based on Szeląg et al. 2016).

due to their negligible impact on the total variance of the model. The model parameters were identified for two catchment inflow hydrographs after intensive rainfall recorded in the summer months (July, September). The effectiveness of the model analysed for the stormwater flow and depth observed in the spring and summer months (May, July), based on the results of measurements of total rainfall depth (P_c), rainfall duration (t_d), mean rainfall intensity ($q = 166.7 \cdot P_c \cdot t_d^{-1}$), and catchment inflow hydrograph volume (V_c) is shown in Table 8.1.

On the basis of the results of the SWMM calculations, the r mean square deviation of the model and the values of the likelihood function (M) were determined. With the likelihood function, the *a priori* rectangular parameter distribution was transformed into the multivariate *a posteriori* distribution. Figure 8.3 shows the variability of the M values (minimum, maximum, mean, standard deviation) for the considered ranges of parameters (ε, d_{imp}, n_{imp}, N_{sewer}, α). The results provided information on the areas in the parameter space with the highest reliability and complemented the sensitivity analysis.

In the case of the flow path width and the channel wall roughness coefficient, the maximum and mean values of the likelihood function in the analysed parameter ranges are very similar, as they vary in the range of 1.55 ÷ 1.58 and 1.18 ÷ 1.30 as well as 1.57 ÷ 1.60 and 1.20 ÷ 1.25. In turn, the minimum values of the likelihood function differ considerably. Given these findings, it can be concluded that the calculation results indicate no clear impact of the considered parameters on the goodness-of-fit of the simulated and measured hydrographs. Moreover, the relationships observed confirm the low values of sensitivity indices for these variables. In the case of the roughness coefficient of impervious areas and the retention and proportion of impervious areas, the values of the likelihood function are equal to the maximum values ($d_{imp} = 2.60$ mm; $n_{imp} = 0.025\,m^{-1/3} \cdot s$; $\alpha = 1.05 \cdot \%Imp_m$) ensuring the best fit of measured and simulated hydrographs. As shown above, these results are also reflected in the high values of the indices of overall and interaction-related sensitivity (Figure 8.2). Next, based on the determined *a posteriori* model output distribution, catchment inflow hydrographs were calculated for the 95% confidence interval (Figure 8.4). In the case of the rainfall events on July 8th, 2011 and May 30th, 2011, the highest uncertainty was observed in the simulated inflow hydrographs. This results from the highest sensitivity of the hydrodynamic model to catchment parameters in the filling

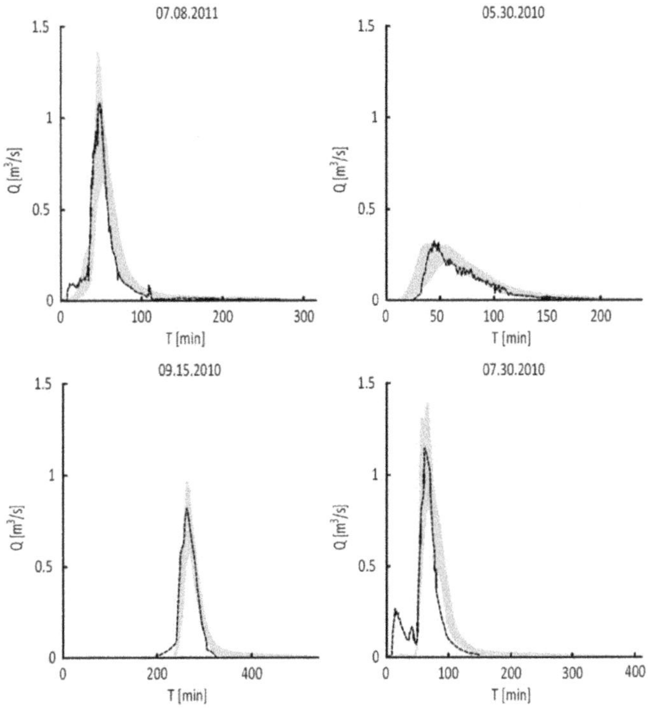

Figure 8.4 Comparison of outflow hydrograph and measurements (Based on Szeląg et al. 2016).

phase. In turn, the lowest uncertainty was obtained for the inflow hydrograph from September 15th, 2010. Considering the calculation results (Figure 8.4), it can be concluded that the model predicts the inflow hydrograph correctly.

The simulation results obtained in this way show that the coefficient values provided by the calibration are subject to considerable uncertainty, which has a significant impact on the results of calculated variability of stormwater runoff from the urban catchment. This aspect is highly important, as the $Q(t)$ values obtained in this way are the basis for modelling the operation of stormwater and combined sewer systems and their facilities, i.e. retention tanks, storm overflows, etc.

8.2 Case studies to manage CSOs

8.2.1 *Rio Vallescura basin*

In combined sewer systems, municipal wastewater and rainwater are collected and carried to wastewater treatment plants (WWTP) in a single network. In the event of intense rainfall, the flow of this line might exceed the capacity of the sewer system, and consequently the excess flow, so-called CSO, is directly discharged into water bodies (Botturi et al. 2020). Contaminations due to CSO events can originate from the dilution of sewage by rainwater, internal contribution by in-sewer sediment re-suspension, and external contribution by runoff (Madoux-Humery et al. 2013). Pollution loads in CSOs are mainly characterised by solids, organic matter, nutrients, metals, organic compounds and pathogenic microorganisms (Montserrat et al. 2013).

Hydrodynamic and water quality models in streams have been extensively applied in different domains and scales to analyse the impact of CSOs in urban streams (Quijano et al. 2017; Riechel et al. 2016; Taghipour et al. 2019). Most of the studies regarding the modelling and/or controlling of CSOs focus on the indicator bacteria such as *E. coli* and intestinal enterococci, since they are crucial indicator parameters for faecal matter and stated in the EC Bathing Water Directive (2006/7/EC). Stormwater system modelling has been used in several studies to achieve greater control over the management of CSOs, applying mitigation measures that limit their impact on the environment. There are several examples where the implementation of sensor activity and the construction of models was applied to monitor the quality of the water system, identifying impact conditions and optimal mitigation solutions. Several studies have been carried out in Germany; an example is represented by the city of Berlin, for which a planning tool has been developed as part of the MIA-CSO research project (Uldack et al. 2013) which integrated a model of the "Info Works CS" sewerage networks, with a model of water quality of the "Hidrax / QSim" rivers and with an impact analysis tool. In this way, it was possible to evaluate the effect of the different boundary conditions of the CSOs, so as to identify the optimal mitigation strategy. Successful examples are also related to the monitoring of bathing water quality, such as the two models developed in Denmark (Bathing Water Forecast System – (Kaas, Erichsen & Roberts)) and in Germany (Flusshygiene – https://bmbf.nawam-rewam.de/projekt /flusshygiene/) which have introduced systems for forecasting the quality of bathing water by monitoring pathogens, such as *Escherichia coli* and Enterococci. Both tools are equipped with systems that allow sharing information with citizens in real time, including possible alarms. Another example is the one of Rimini, in Italy, where the role of modelling was fundamental in the implementation of a Bathing Safeguard Plan, which allowed identifying a series of necessary structural interventions. Specifically, interventions were made to adapt the treatment systems and the collection systems, to separate the networks and to create storage volumes. In addition to this, the methodology implemented allows the municipality to adopt preventive management measures based on predictive information of the water bathing conditions.

In Central Italy a methodological approach has been applied at different basins to quantify both qualitatively and quantitatively the impact of CSO discharge. In particular, the example reported is the study performed in Vallescura Basin. Rio Vallescura is a small-medium canal located in the municipality of Porto San Giorgio, in Central Italy. The drainage system works by gravity, without any use of pumping, and conveys the sewer into a principal collector placed to Rio Vallescura, in which secondary pipelines arriving from the nearby hills are also connected and ends up in Lido di Fermo WWTP. The network is characterised by the presence of 35 spillways discharging into Rio Vallescura (Figure 8.5). The study integrated and further experimentally validated dynamic rainfall-runoff simulation and water quality models to create a CSO pattern based on different rainfall scenarios and pollutant loads. Most critical spillways were identified that cause most of the faecal bacteria contamination and loss of excellent bathing water quality in Porto San Giorgio. On the basis of the simulations, chemical and physical disinfection methods were proposed and in situ and ex situ tested in order to eliminate microbial contamination. Then, possible mitigation scenarios were developed for the most critical spillways to minimise their impact on the receiving water body.

An SWMM model of the catchment was created by implementing the urban drainage network, which included the setting of hydraulic structure (i.e. pipes dimension) as well as the knowledge of basin hydrological characteristics (Figure 8.6). Both quantitative and qualitative assessment and calibration were considered. In a preliminary phase, the quantitative analysis was conducted considering both dry and wet periods. A flow meter was placed at point S282

Figure 8.5 Rio Vallescura basin. (Based on: Crocetti et al. 2021.)

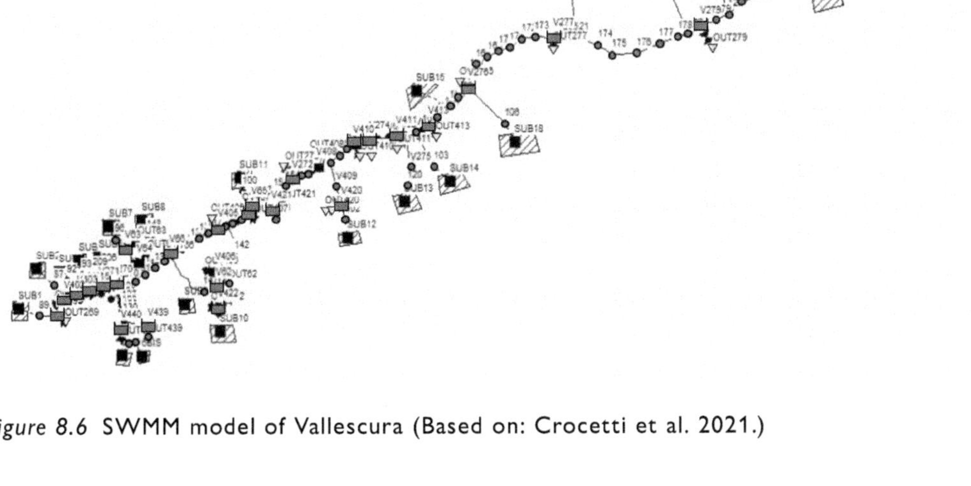

Figure 8.6 SWMM model of Vallescura (Based on: Crocetti et al. 2021.)

(Figure 8.5) to provide hydraulic influent daily flow into the sewage system. An average value of 1,506 m^3/d was obtained from the specific elaboration of the data meter during the period considered (from 4th June to 12th June 2019), which identified a specific influent flow value of 136 L per person per day. The latter provided the total discharged flow by 7,227 PE, inserted as input in SWMM. Calibration of the model was performed both in dry and wet periods by comparing the measured values to the results extrapolated by SWMM.

The qualitative assessment was supported by a sampling campaign in five key points (S272, W409, S276, S280, S282) in the sewer. At each point, the concentrations of the main pollutants (COD, TKN, TSS, BOD_5, *E.coli*) have been measured and given as input to the model for the four most upstream points (S272, W409, S276, S280). The calibration was subsequently performed, checking that the model output in the downstream sampled point (S282) could be compared to the laboratory analysis values. The percent bias (PBIAS) was calculated for this purpose to compare the simulated data with the observed data, providing information about overestimation (positive PBIAS) and underestimation (negative PBIAS) of the calibrated model (Gupta et al. 1999). The final values obtained for COD and TKN are equal to 5.06% and 7.30%, a rating performance lower than 10%, which can be considered "very good" (Moriasi et al. 2007).

After the calibration of the model, different simulation scenarios were considered in order to analyse the impact of the CSOs during rainfall events of different intensities. Three return times (RT) were considered, representing the probability that a harmful event occurs in each area within a specified period. These rainfall heights are equal to 12.8 mm for RT = 1 y, 41 mm for RT = 5 y and 48.7 mm for RT = 10 y. To have a continuous rainfall scenario, a "typical year" was calculated through a statistical evaluation. Accordingly, all the rainfall data from 1999 to 2019 were analysed to identify the most representative year that deviated less than the average value of the period considered (Crocetti et al. 2021). A specific evaluation to consider the seasonal variability of rainfall was performed, analysing separately winter and summer periods. Typical concentrations of runoff water were evaluated and given as input to the SWMM model. Finally, the impact was also considered of stormwater runoff linked to dwellings that are not connected to the public sewage system (4% of the total users in the basin) was considered and modelled (Crocetti et al. 2021).

In conclusion, the source of environmental impact considered in the study were three:

1 CSOs loads,
2 Load of the users not connected to the sewer system,
3 WWTP effluent.

Through specific simulation of the SWMM model it was also possible to identify the origin of the contamination inside the sewer, distinguishing between the discharge of the utilities connected to the sewer and the contribution of the runoff of rainwater.

Results of the simulations allowed, firstly, verifying the compliance with regional regulations in terms of overflows by calculating the "P ratio" which is the permitted ratio between peak flow in rainy weather and average dry flow, which must be higher than 4. This condition was fulfilled for all the spillways and for all the meteorological events considered.

In a second stage, before the direct analysis of the impacts, the flooding of the system was considered since, especially for high intensity events (RT of 5 y and 10 y), this condition involved more than 40% of the nodes in the model. This means that for strong events (5 y and 10 y of RTs), the sewer is not sufficient and, to account this, the flooded volumes must be considered as an overflow.

Finally, including the influence of flooding, the results of the simulations highlighted the activation of a maximum of 25 in 35 spillways for the strongest events. Considering that in the typical year there were about 201 rain events (including minor events), there were several activations of the spillways, ranging from a minimum of once to a maximum of 69 times per year, with an average of 24 active spillways per year (Crocetti et al. 2021). The focus of the study was related to faecal bacteria contamination, which could affect the quality of the bathing water of Porto San Giorgio. What emerged from the study is that E. coli mainly originated from the discharge of the utilities connected to the sewer (10^{14} UFC/d), even if the value associated with the runoff of the rainwater turned out to be still high (10^{11} UFC/d). In Figure 8.7, the total flow and pollutant concentration in the overflow based on the SWMM model with a typical year simulation are reported.

In terms of environmental impact, considering the three contributions already introduced (Figure 8.8), it can be observed that the load associated with the users who were not connected to the sewer system turned out to be much smaller than the overflow, less than 4% for all contaminants, especially regarding the E. coli that had an impact less than 1%. The contribution of BOD_5 on the environment was associated with the WWTP effluent (55%) as well as the overflows (42%), and the same for COD, with 62% for effluent and 36% for spillways activation. E. coli and TKN, on the other hand, had different origins; in fact, E. coli was derived almost entirely from CSOs (99.2%), while TKN was associated mainly with the discharge of the effluent (95%). Finally, TSS was mainly associated with CSOs (69%), but also had a contribution from the WWTP effluent (29%) (Crocetti et al. 2021).

From the analysis performed, the most harmful spillways have been identified and considered for the definition of specific optimisation scenarios. In particular, the critical spillways are: S277 (25.3% of volume and 29% of the E. coli load); S279 (9.1% of volume and 8.4% of the E. coli load); S280 (8.5% of volume and 15.5% of the E. coli load);, and S283 (13% of volume and 33.8% of the E. coli load).

The minimisation scenarios had the main objective of reducing the E. Coli release in surface water by reducing the quantity and/or quality of the overflows. For this reason, two different actions have been evaluated:

1 The possibility to treat at the WWTP part of CSOs in order to reduce the quantity of flows discharged.
2 The possibility to locally treat the overflows through disinfection before the discharge.

Two scenarios were subsequently identified: in the first one, wastewater from S283 (the last spillway before the WWTP) is treated at the WWTP up to the maximum capacity available, while the Scenario 2 includes both Scenario 1 and in situ treatment at the two most critical spillways (S280 and S277).

To verify the feasibility and the performance of the two scenarios, different disinfection tests were performed both in the laboratory and pilot scale. Four disinfection methods were considered: performic acid (PFA), peracetic acid (PAA), ozone, and UV treatment. The three chemical

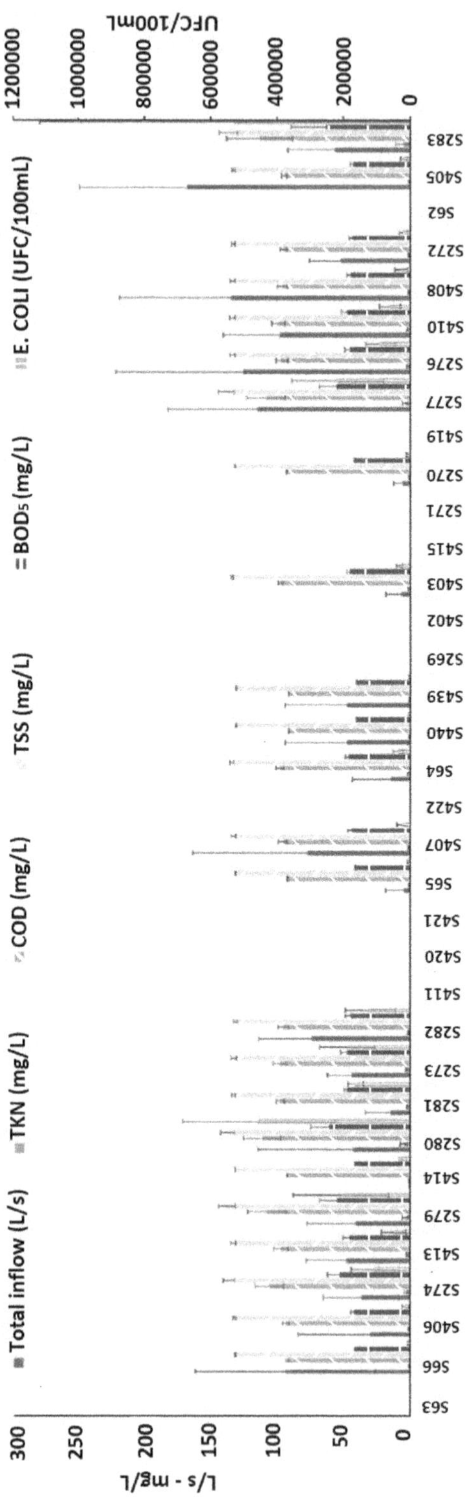

Figure 8.7 Total flow and pollutant concentration in the overflow based on the SWMM model (typical year simulation). (Based on: Crocetti et al. 2021.)

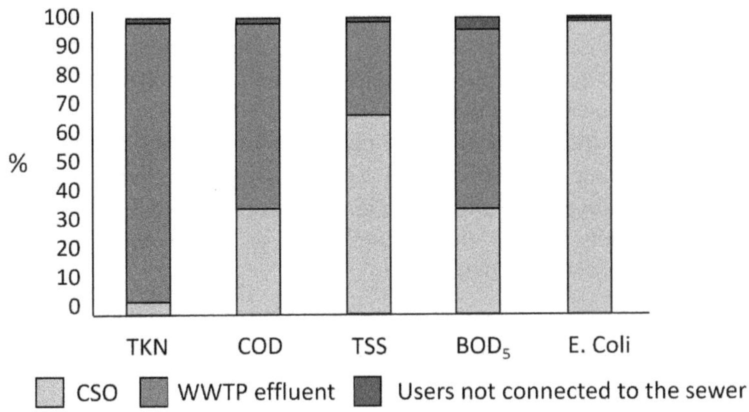

Figure 8.8 Environmental impact. (Based on: Crocetti et al. 2021.)

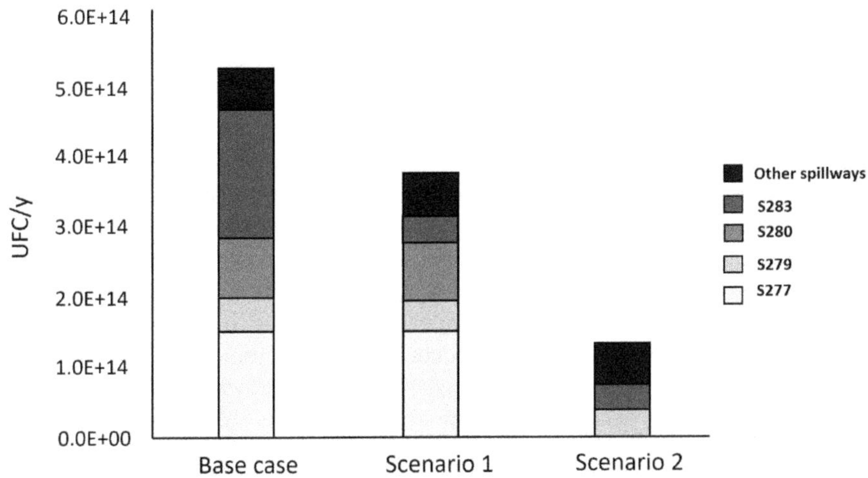

Figure 8.9 Optimisation scenarios. (Based on: Crocetti et al. 2021.)

treatments were tested in laboratory, while the UV was tested on-site using two filtration units in parallel followed by a small UV system to remove TSS and to increase UV transmittance. The disinfection performances obtained with laboratory and pilot scale tests, have been applied to the flows resulting from SWMM simulations to evaluate the impact of the different mitigation scenarios. In addition, for Scenario 1, a screening of the volumes available at the WWTP was done to quantify the specific amount of wastewater which could be delivered to the WWTP instead of being discharged into Rio Vallescura.

The final results of the optimisation scenarios compared with the baseline condition are represented in Figure 8.9.

This case study represents an example of application of the SWMM model to quantify the impact of CSOs and identify possible solutions to reduce this impact and, consequently increase the bathing water quality. To implement such a study, the qualitative and quantitative knowledge of real drainage flows is fundamental. This involves data acquisition and specific measurements on the catchment, on the infrastructure and on the collected flows, which are often not easily available. In many cases, this phase also require specific campaigns for flow and concentration measurement. Only in this way it is possible to have a validated model and, therefore quantify the impacts and take action to mitigate them.

The simulated data and the results achieved represent fundamental knowledge of the quantitative and qualitative impact of network spillways, allowing planning, under different conditions, priority of interventions and / or the creation of a targeted monitoring network of critical spillways. Validated models identify the infrastructural parts to support the development of more complex integrated environmental models.

8.2.2 *Model case studies Hron and Drava*

8.2.2.1 *Introduction*

Urban sewerage networks in both countries (Slovakia and Croatia, especially in larger cities) were very often designed as sewerage networks of combined type, i.e. mixing of stormwater and wastewater is assumed. These systems suffer very often from absence of centralised or decentralised retention and detention elements, thus from hydraulic point of view, in case of storm events it is necessary to discharge a large part of the stormwater through CSOs directly to the receiving water. This approach, of course, leads to a deterioration of water quality in receiving water bodies.

The methods mostly used for evaluating the influence of wastewater on the receiving water are still the imission and emission methods. Evaluation of the influence of the wastewater on the receiving water can be done in various ways: beside the classical approach (volume, concentration limits, emission, imission standards or limits), are now coming into foreground biological methods, evaluation directly the influences on the aquatic life. Both methods have been known for a long time and are worked out in detail in the literature and also in the legislation, where they are successfully used for evaluating the influence of the continuous outflow of waste water – from the communal or industrial waste sewerage plants etc. However, the question is how to evaluate the effects of discontinuous sources of pollution, for example, of the CSOs on individual sewage networks?

The general goal is to achieve at least the "good" water quality status of the entire water body. The basic approach towards solving the effect that wastewater has on the receiving water is based on the achieving the required quality of the receiving water and preventing the discharge of wastewater in large amounts, according to what the receiving water is used for (fish protected water bodies, bathing, drinking water etc.). It is necessary to follow the water quality criteria for surface water, in order to ensure required limits defined in the legislation, for example a minimal concentration of dissolved oxygen in the receiving water, the maximal concentration of harmful substances, e.g. ammonia nitrogen (NH_3–N) in reference to the pH and temperature of the water in the receiving water. Such requirements should be based on environmental limits, e.g. respect the values LC_x with prescribed frequency, where the x is the required safety level. LC_{50} means 50% death rate of fish with the stated pollutant concentration and exposition time).

Such regulation method could be called "imission" regulation method, and can be used also for non-continuous (discontinuous) pollution sources, like CSOs.

8.2.2.2 *Model data*

For the mathematical simulation of the water quality, the MIKE 11 model (Danish Hydraulic Institute, DHI) was used. Physical definition of the river beds was made by the graphic editor of the simulation software MIKE 11. As a basis, digital maps in DWG format were used, obtained from the river authorities, eventually from other sources (Mišík 1998; SHMU, 1999–2009; Bara & Velísková 2012; Bara et al. 2012). Into this map base, all the river bed direction points and also all the connection points – in which the tributaries, pollution release, eventually other objects were defined – were entered.

There is not sufficient information available about the flow hydraulics (roughness coefficients) for the modelled river section, as well as for the kinetics of physical, chemical, and biological processes in the river (dispersion coefficients, re-aeration or de-oxygenation coefficients etc.). These data are required by the water quality module. Individual coefficients, affecting the process kinetics, were set up according to the results of the steady-state simulation. Results of these simulations, computed in the final section of the modelled river section, should correspond to the monitored water quality values. Also, there were no direct measurements of pollution concentrations of the CSO effluent available, so the data obtained during the survey in Slovakia were used (Sztruhár et al. 2002). These values, used for the simulation, are presented in Table 8.2.

For the water quality simulation, level 5 of the standard model was used, which allows modelling of concentrations of dissolved oxygen (O_2), ammonia nitrogen (NH_4–N), nitrate nitrogen (NO_3–N), biochemical oxygen demand (BOD_5) and phosphorus (P) in dissolved and particulate form, including a delayed oxygen consumption. This level of complexity comprises modelling of re-aeration, degradation of organic matter - instantaneous oxygen consumption, nitrification and denitrification processes.

8.2.2.2.1 CASE 1: DRAVA RIVER (CROATIA)

The part of the Drava River from rkm 69.118 (downstream from the monitoring profile no. 29111 between the villages of Donji Miholjac and Podravski Podgajci) to the mouth of the Danube River (rkm 0,000) was modelled. In this section of the Drava River, the river bed diverts from state border with Hungary and flows through the territory of Croatia. In this section the Drava River has a character of meandering flow with lots of side branches, but no major tributaries.

Water quality models for rivers are usually used to resolve or predict adverse conditions on the river, which mainly occur at relatively low discharges. The simulation for this case study was

Table 8.2 Output concentration values of the modelled pollution parameters – outflow from CSO Osijek (Sztruhár et al. 2002)

Indicator	*Unit*	*Value*
Dissolved oxygen O_2	mg/L	2
Water temperature	°C	25
Ammonia nitrogen (NH_4–N)	mg/L	6.21
Nitrate nitrogen (NO_3–N)	mg/L	1.28
Dissolved phosphorus – P	mg/L	2.37
Particular phosphorus – P	mg/L	0.26
Biochemical oxygen demand – BOD_5	mg/L	175

considered with quasi-steady-state discharge near to the minimum discharge $Q_{min} = 152\,m^3/s$. This flow rate was defined as the upper boundary condition at 69.118 rkm profile. The lower boundary condition was defined as the constant water level 79 m above sea level (the mouth of the river Drava and Danube in rkm 0.543).

For steady state, the 90-percentile of water quality (c_{90}) values from the monitoring station Donji Miholjac (monitoring profile no. 29111, see Figure 8.10) were used, i.e. the values of the quality indicators corresponding to 90% quantile of not exceeding probability (for oxygen O_2, the 10% percentile) the substance concentration in the river.

In this scenario, the effect of wastewater, relieved through CSOs from the sewer network of the City of Osijek was simulated. The city has a total of 8 CSOs flowing to the Drava River; for simplification reasons, the effluent was concentrated into three points. The data from the work (Krsnik 2011) were used for this simulation scenario. In this diploma work, is the presented stormwater discharges monitoring in the sewer network of the Osijek city, as well as statistical evaluation of rain gauge data from GMS Osijek station, based on historical rain gauge data series recorded in 1959–1991.

Considering the fact that the raw wastewater flow from CSOs depends on the rainfall intensity, several options were simulated with different frequency and duration of the storm events.

The total catchment area of the sewer system is 7.56 km², the peak runoff coefficient by (Krsnik 2011) was considered, using the value $\psi = 0.36$. The CSO effluent discharge time course was not simulated, the outfall duration was considered equivalent to the storm duration. The CSO effluent discharge was also considered constant, simple intensity-area method was used for its determination.

8.2.2.2.2 CASE 2: HRON RIVER (SLOVAKIA)

The first part of the project was modelling the stormwater runoff from the sewer system of the city Banská Bystrica in aim to simulate CSO wastewater discharges from CSO structures to the river Hron. Block rainfalls with the periodicity of 0.033, 0.1, 0.2, 0.5, 1.0 and 5.0 with duration of 15, 30, 60 and 180 minutes were used.

The Banská Bystrica town operates a combined sewer system conveying both storm and sanitary wastewater into one central WWTP located downstream of the town and near the Hron

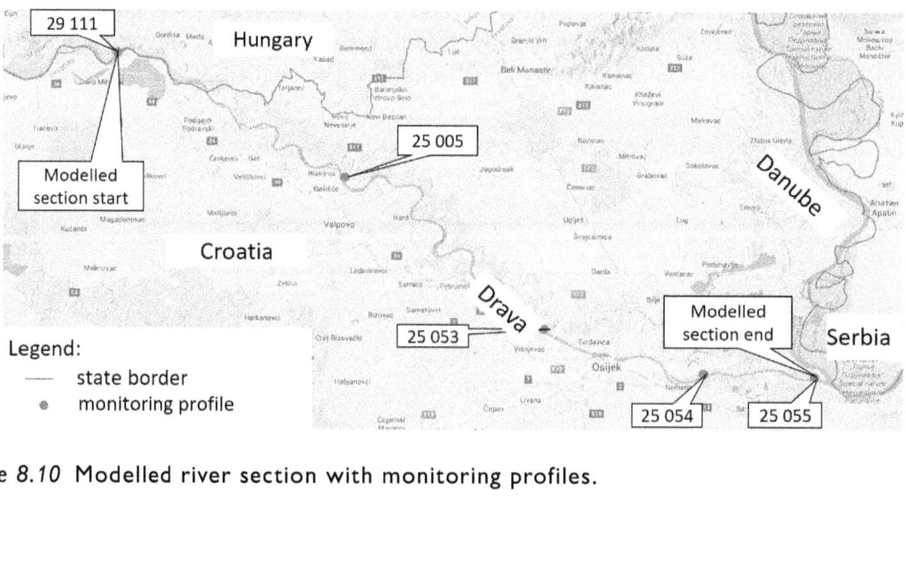

Figure 8.10 Modelled river section with monitoring profiles.

River. In the town itself, a combined sewer system is built; however, some parts of the town are drained by a stormwater system discharging the stormwater directly into the recipient (the Hron River or local recipients). The new town development areas, eventually the municipalities located nearby, are drained by a sanitary sewer system.

The total length of the simulated sewer network (main sewer line, collectors and the street sewers) is 146.893 km long. The length of the main collector A (see Figure 8.11) is approximately 10.16 km with the diameter of 1,000 to 1,800 mm.

As an input for the model of sewer network digitalised maps were used. Digital map was used as for localisation of nodes and segments of the sewer network, as well as for interactive editing of sewer catchment data (covering partial areas of the sewer lines) and automatic calculation of the area, runoff coefficients and other hydrologic parameters. The surface of parts of urbanised

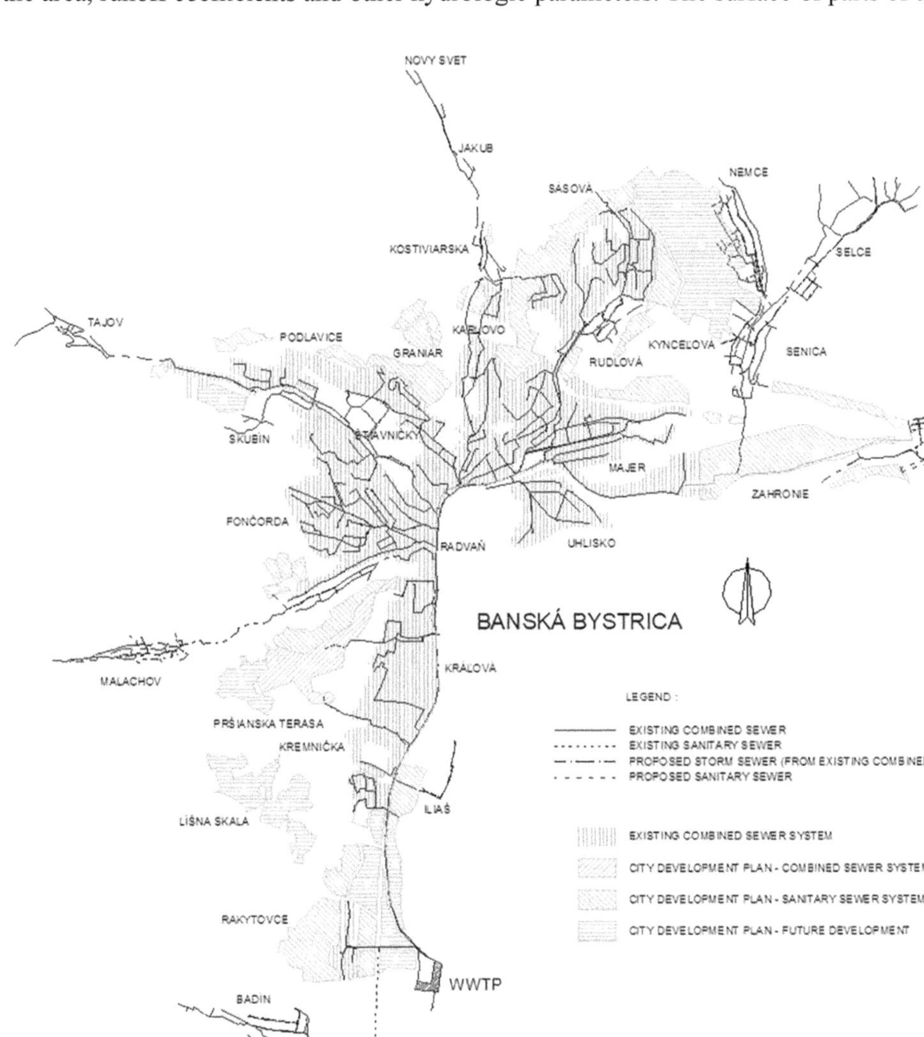

Figure 8.11 Schematic layout of sewer network of the Banská Bystrica town.

areas are described as impervious (roofs and roads) and pervious (for example green areas). The impervious surfaces were defined as closed polygons and they were automatically stored in the database, as well as the other boundary of sewer network parts - elementary catchment areas.

The program checks the data accuracy using the user visual check of the geographical data and numerical check of the data – minimum and maximum data limits as well as the logical data check. Automatic check of topology data is also carried out.

Complete database of the sewer network input was then transferred by a built-in routine into the simulation model MOUSE (DHI 2000).

The main data of the simulated sewer network of the Banská Bystrica town are shown below:

Total drainage area of the urban catchment	1,209.367 hectares
Total reduced drainage area of the catchment	302.68 hectares
Average runoff coefficient:	0.25
Number of elementary catchments:	963
Minimum ground elevation:	319.5 m above sea level (WWTP)
Maximum ground elevation:	495.0 m above sea level
Number of inhabitants:	84,919
Number of nodes (structures):	1,288
Number of CSOs:	29
Minimum pipe bottom elevation:	318.0 m above sea level (WWTP)
Maximum pipe bottom elevation:	492.90 m above sea level
Number of outfalls from sewer network:	25
Number of sections of sewer network:	1,291
Length of sewers:	146.893 km
Volume of sewers (total):	50,789 m^3

The difference in ground elevations between the centre town and outskirt town parts is more than 160 m. The schematic of the sewer network of the Banská Bystrica town is shown in Figure 8.11.

The modelling was performed using block rainfalls with constant intensity of various frequencies and durations from 15 to 180 minutes.

Overall, 4 basic alternatives of the stormwater management were simulated on the city catchment (according the SK legislation – please note that the sequence numbers of alternatives are used for marking the IDF curves in consequent figures):

1 Dilution (mixing) ratio of the CSO 1:4
2 Dilution (mixing) ratio of the CSO 1:8
3 Storm tank with the volume necessary for accumulation of CSO volume with the frequency $p = 0.5$ (2 years return period)
4 Storm tank with the volume necessary for accumulation of CSO volume with the frequency $p = 5$ (return period 5 times in a year)

The second part of the case study was to create a mathematical model of the receiving water – the river Hron to simulate the water quality and its changes due to CSOs from the city of Banská Bystrica. The receiving water – river Hron was modelled in the space extent from the river kilometre 261.300 (Valkovňa profile), i.e. from the first monitored profile on the river to the estuary (km 0.000), which is practically the whole river length except for a short section near the spring (see Figure 8.12).

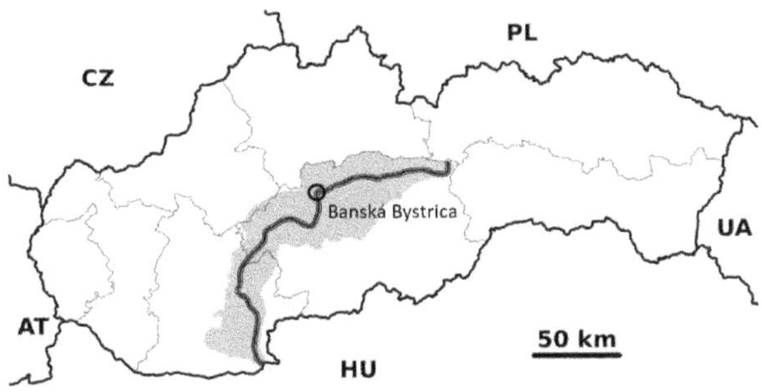

Figure 8.12 Hron river, its river basin (grey) and the city of Banská Bystrica on the Slovakia outline map.

Simulations for two "basic" states of hydrologic situation at the river Hron were modelled – the long-term average flow (Q_a – for the river Hron, and also in all the tributaries) and flow, which are close to the minimal flows –for a 355-day flow Q_{355}.

In the first alternative (average flow Q_a) the average pollution concentrations to simulate the water quality were used; in the second alternative (Q_{355}) the simulation was performed using the value c_{90} (90% distribution percentile of the most unfavourable concentration value). As input values, the information gained from the annual water quality report (SHMÚ) was used.

The modelled flows and the quality of the wastewater discharged from CSOs were used as input to the model of the water quality in the river Hron. In first step, only the alternative was evaluated, where the discharged wastewater will be discharged into the average river flow (Q_a), but the alternative that the discharged wastewater will be discharged into the 355-day flow (Q_{355}) was also considered.

8.2.2.3 Model results

8.2.2.3.1 CASE 1: DRAVA RIVER (CROATIA)

In this scenario, discontinuous emission of the waste water to the Drava river has been simulated, which is very common in practice – hydraulic overflow of sewage from sewer systems during storm events through CSOs. This is a relatively short event, but the more intense hydraulic and pollution load of the recipient, depending on the intensity and duration of the storm event.

All used storm events had different duration and different frequencies. Simulation results are documented in Figures 8.13 and 8.14. Graph in Figure 8.13 documents the pollution – BOD_5 concentration longitudinal profile, using the 30-minute storm event with return period 10 years ($p = 0.1$). The figure shows only a section of the river Drava, downstream of the Osijek city, from rkm 24.0 to the estuary to the Danube (rkm. 0.0). As it can be seen in the figure, short-term intense rain caused higher pollution concentration, but only with short-term duration, whereas the longer storm event causes smaller increase of the pollutants concentrations, but with longer effect time.

These different effects of specific pollutants on the receiving water quality can be expressed with IDF curves (Intensity – Duration – Frequency) (Sokáč 2010). To illustrate the simulation results, an example is shown in Figure 8.14 - IDF curves of the concentration and effect duration of ammonia nitrogen NH_4-N. All curves are simulation results using storm events with different duration and return period of 10 years.

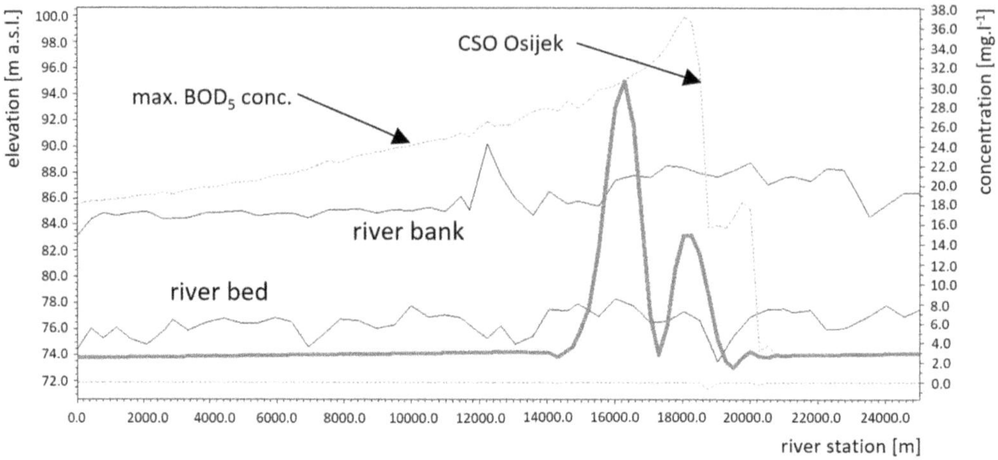

Figure 8.13 Modelled BOD_5 concentrations in Drava river (rain duration 30 minute, return period 10 years).

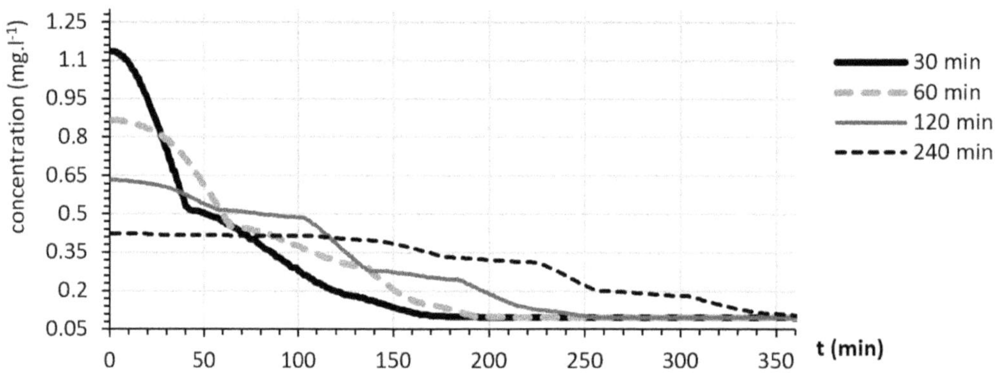

Figure 8.14 IDF curve of ammonia nitrogen (NH_4–N) concentrations in the Drava river (return period 10 years, river km. 17.825).

8.2.2.3.2 CASE 2: HRON RIVER (SLOVAKIA)

Simulation results of the water quality are shown in form of the IDF (Intensity – Duration – Frequency) curves for the basic water quality parameters - the dissolved oxygen concentration, also for BOD_5 and N–NH_4. IDF curves for various alternatives are displayed in Figure 8.15. These curves show the intensity of the negative effects and the duration of this exposition on the biocoenosis of the river. The numbers in the legend represent the stormwater management alternative.

As shown in the figures, the discharging of wastewater represents a relatively short-term stress on the receiving water. It is necessary to mention that the catchment of Banská Bystrica is relatively specific – it has relatively large slopes and a large number of CSO structures, which means that large volumes of the wastewater infiltrate into the receiving water through the CSO structures very fast (very small retention capacity of the system, a total absence of accumulation structures in the system).

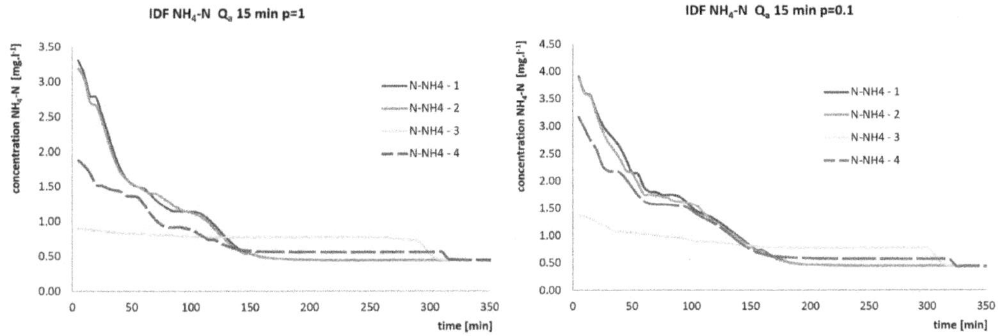

(a) (b)

Figure 8.15 IDF lines for the ammonium nitrogen (N–NH_4) concentration in the receiving water during a 15-minute rain event with a periodicity of occurrence p = 0.1 and 1.0.

While modelling the water quality in the Hron river, the conclusion was drawn that the pollution concentration in the river Hron is not decreased only by the self-cleaning ability of the river, but also by the hydrology and the hydraulics of the stream, which is the dilution, mixing and dispersion. For example, after the CSO event, due to the lateral inflows (tributaries) of Hron, discharge in river doubles within a short river section (from profile Sliač to profile Žarnovica, approx. 60 km), due to this fact the concentration decreases to one half of the original concentration. This fact has a significant effect on the modelling of the concentration of dissolved oxygen – water with a large amount of oxygen flows from the tributaries to the river Hron (mountain areas, tributaries with large slope), which significantly improves the oxygen balance of the Hron river after the CSO event. This situation is relatively specific, even though it is possible to assume it on a large number of rivers and streams in Slovakia.

The results of the study also show the effect of storm tanks in the sewage system of the urbanised catchment. The obtained result is that when a storm tank is included into a sewer system, the pollution concentration in the receiving water can be three to four times lower than in network schemes without storm tanks.

When comparing individual methods of stormwater management (see Figure 8.15), the following facts can be stated:

1 The size of the mixing ratio for short-term intensive rain events practically does not have any effect on the quality of water in the receiving water,
2 The most effective method of structural regulation is, despite of many disagreements and problems, the construction of storm tanks.

Ad 1: the judged alternatives represent minimum and maximum requirements according to the current Slovak legislation. As it can be seen in Figures 8.14 and 8.15, during short-term rain events with a relatively low periodicity of occurrence, there is not a great difference between these two alternatives concerning the quality of water in the receiving water. The difference is shown more during longer rain events or during heavy rains with lower periodicity. Also, when using a higher mixing ratio, the periodicity of occurrence of the CSO events (discharges of the raw wastewater into the receiving water) decreases, which in some cases can have a significant importance.

Ad 2: this conclusion is not surprising in any way; as these results were expected. Nevertheless, they are shown here, because they easily show the effect of storm tanks on the quality of water in the receiving water – the concentration of harmful substances in the receiving water in the described cases is three to four times lower (alt. 3) than alternatives without storm tanks.

8.2.2.4 Summary of analysis concerning CSOs functioning

This chapter presents two modelling studies, focused on stormwater runoff modelling from urbanised catchments through sewer system to the receiving water and analyses its effect on water quality in the receiving water.

Among other, the second case study (Hron river, Slovakia) analysed the possible strategies of stormwater management in urban catchments within the present legislative framework in Slovakia.

In both case studies, the pollution load on receiving water and its water quality deterioration takes relatively short time, so for the impact analysis the IDF (intensity – duration – frequency) curves are used, reflecting the short-term deterioration of water quality. The IDF curves are presented for all scenarios of urbanised catchment stormwater management.

The results of presented model study point out to the limitations of the current legislation approach, as well as the effectiveness of different stormwater management measures (scenarios) for urbanised catchments in terms of their impact on water quality on the receiving water. Results also show that such modelling approach can be very useful for modelling of impacts on the water quality in receiving water and can be also very helpful in evaluation of various strategies and management practices focused on the receiving water quality improvement.

Despite all the stated difficulties and problems, the authors think that using the imission method for evaluating the influence of wastewater outflows from CSO is possible. The remaining question is, how can it be applied in practice and under the conditions, where the mathematical models are available, but there is a significant lack of the necessary data. In both presented studies was statistical approach used, where the average data (or specific percentile data) were used for hydraulic and qualitative parameter specification in receiving water, as well as for the urban catchment effluent from CSOs.

References

Abdellatif, M., Atherton, W., Alkhaddar, R.M. & Osman, Y.Z. 2015. Quantitative assessment of sewer overflow performance with climate change in northwest England. *Hydrol. Sci. J.*, 60(4): 636–650.

Abt, S.R. & Grigg, N.S. 1978. An approximate method for Sizing Detention Reservoirs. *J. Am. Water Resour. Assoc.*, 14(4): 956–965.

Acharya, N., Frei, A., Chen, J., DeCristofaro, L. & Owens, E.M. 2017. Evaluating stochastic precipitation generators for climate change impact studies of New York City's primary water supply. *J. Hydrometeorol.*, 18(3): 879–896.

Acquaotta, F., Frantianni, S. & Venema, V. 2016. Assessment of parallel precipitation measurements networks in Piedmont, Italy. *Int. J. Climatol.*, 36(12): 3963–3974.

Addison-Atkinson, W., Chen, A.S., Memon, F.A. & Chang, T.J. 2022. Modelling urban sewer flooding and quantitative microbial risk assessment: A critical review. *J. Flood Risk Manag.*, 15: e12844.

Akan, O.A. 1985. Kinematic-wave method for peak runoff estimates. *J. Transp. Eng.*, 111(4): 419–425.

Akan, O.A. 1990. Single outlet detention pond analysis and design. *J. Irrig. Drain. Eng.*, 116(4): 527–536.

Akan, O.A. & Houghtalen, R. 2003. *Urban Hydrology, Hydraulics, and Stormwater Quality. Engineering Applications and Computer Modeling.* John Wiley & Sons, Hoboken, NJ.

Andrés-Doménech, I., Montanari, A. & Marco, J.B. 2010a. Stochastic rainfall analysis for storm tank performance evaluation. *Hydrol. Earth Syst. Sci.*, 14(7): 1221–1232.

Andrés-Doménech, I., Múnera, J.C., Francés, F. & Marco, J.B. 2010b. Coupling urban event-based and catchment continuous modelling for combined sewer overflow river impact assessment. *Hydrol. Earth Syst. Sci.*, 14(10), 2057–2072.

Annen, G. & Londong, D. 1960. *Vergleichender Betrachtung zur Bemssung von Ruckhaltebecken.* Techn. Wiss. Mitt. d. Emshergenossenschart, Essen, Germany, p. 3.

Archer, G., Saltelli, A. & Sobol, I.M. 1997. Sensitivity measures, Anova-like techniques and the use of bootstrap. *J. Stat. Comput Simul.*, 58(2): 99–120.

Arnbjerg-Nielsen, K., Willems, P., Olsson, J., Beecham, S., Pathirana, A., Bülow Gregersen, I., Madsen, H. & Nguyen, V.T.V. 2013. Impacts of climate change on rainfall extremes and urban drainage systems: A review. *Water Sci. Technol.*, 68(1): 16–28.

Arnell, N.W. 2011. Uncertainty in the relationship between climate forcing and hydrological response in catchments. *Hydrol. Earth Syst. Sci.*, 15(4): 897–912.

Aron, G. & Kibler, D.F. 1990. Pond sizing for rational formula hydrographs. *J. Am. Water Resour. Assoc.*, 26(2): 255–258.

As-syakur, A.R., Adnyana, I.W.S., Arthana, I.W. & Nuarsa, I.W. 2012. Enhanced Built-Up and Bareness Index (EBBI) for mapping built-up and bare land in an urban area. *Remote Sens.*, 4(10): 2957–2970.

ATV-A 118. 2000. *Hydrauliczne Wymiarowanie Systemów Odwadniających.* Komentarz do ATV-A 118. Wydawnictwo Seidel-Przywecki, Warszawa.

ATV-A 118. 1999. *Hydraulic Dimensioning and Verification of Drainage Systems.* GFA, Hennef.

ATV-DVWK-A110P. 1988. *Wytyczne do hydraulicznego wymiarowania i sprawdzania przepustowości kanałów i przewodów ściekowych.* Wydawnictwo Seidel-Przywecki, Warszawa.

Baek, S.-S., Choi, D.-H., Jung, J.-W., Lee, H.-J., Lee, H., Yoon, K.-S. & Cho, K.H. 2015. Optimizing low impact development (LID) for stormwater runoff treatment in urban area, Korea: Experimental and modeling approach. *Water Res.*, 86(1): 122–131.

Bacchi, B., Balistrocchi, M. & Grossi, G. 2008. Proposal of a semiprobabilistic approach for storage facility design. *Urban Water J.*, 5(3): 195–208.

Baker, W.R. 1979. Stormwater detention design for small drainage areas. *Public Works*, 108(3): 75–79.

Balistrocchi, M. & Bacchi, B. 2011. Modelling the statistical dependence of rainfall event variables through copula functions. *Hydrol. Earth Syst. Sci.*, 15(6): 1959–1977.

Balistrocchi, M., Grossi, G. & Bacchi, B. 2009. An analytical probabilistic model of the quality efficiency of a sewer tank. *Water Resour. Res.*, 45(12): 1–13.

Ballinas-González, H., Alcocer-Yamanaka, V. & Pedrozo-Acuña, A. 2016. Uncertainty analysis in data-scarce urban catchments. *Water*, 8(11): 524.

Banasik, K. 2009. *Predicting the Flood Hydrographs for Small Urban Watersheds* (in Polish). Warsaw University of Life Sciences – SGGW Press, Warsaw.

Barco, J., Wong, K.M. & Stenstrom, M.K. 2008. Automatic calibration of the U.S. EPA SWMM model for a large urban catchment. *J. Hydraul. Eng.*, 134(4): 466–474.

Bartušková, I. 1985. *Problémy výpočtu kanalizačných sítí. Kandidátska dizertačná práca*, VUT Brno, Brno.

Basha, H.A. 1995. Routing equations for detention reservoirs. *J. Hydraul. Eng.*, 121(12): 885–888.

Bates, B.C., Kundzewicz, Z.W., Wu, S. & Palutikof, J.P. 2008. *Climate Change and Water.* Technical Paper of the Intergovernmental Panel on Climate Change. IPCC Secretariat, Geneva, Switzerland.

Becker M., Prinz, R. & Liebig, T. 2001. Effectiveness of a stormwater tank with overflow with a downstream stormwater holding. *KA-Wasserwirtschaft, Abwasser, Abfall*, 1, 16–24.

Behrouz, M.S., Zhu, Z., Matott, L.S. & Rabideau, A.J. 2020. A new tool for automatic calibration of the Storm Water Management Model (SWMM). *J. Hydrol.*, 581: 124436.

Bellos, V. & Tsakiris, G. 2016. A hybrid method for flood simulation in small catchments combining hydrodynamic and hydrological techniques. *J. Hydrol.*, 540: 331–339.

Beven, K. 2001. How far can we go in distributed hydrological modelling? *Hydrol. Earth Syst. Sci.*, 5(1): 1–12.

Beven, K. & Binley, A. 1992.The future of distributed models: Model calibration and uncertainty prediction. *Hydrol. Process.*, 6(3): 279–298.

Bhunya, P. K., Berndtsson, R., Ojha, C.S.P. & Mishra, S.K. 2007. Suitability of gamma, chi-square, Weibull, and beta distributions as synthetic unit hydrographs. *J. Hydrol.*, 334 (1–2): 28–38.

Bin Ali, M.T., Horoshenkov, K.V. & Tait, S.J. 2011. Rapid detection of sewer defects and blockages using acoustic-based instrumentation. *Water Sci. Technol.*, 64(8): 1700–1707.

Blocken, B., Derome, D. & Carmeliet, J. 2013. Rainwater runoff from building facades: A review. *Build Environ.*, 60: 339–361.

Błaszczyk, W., Roman, M. & Stomatello, H. 1974. *Kanalizacja t. 1.* Wydawnictwo Arkady, Warszawa.

Bogdanowicz, E. & Stachy, J. 1998. *Maksymalne Opady Deszczu w Polsce – Charakterystyki Projektowe.* Materiały Badawcze IMGW. Seria Hydrologia i Oceanologia, 23. Instytut Meteorologii i Gospodarki Wodnej, Warszawa.

Botturi, A., Daneshgar, S., Cordioli, A., Foglia, A., Eusebi, A.L. & Fatone, F. 2020. An innovative compact system for advanced treatment of combined sewer overflows (CSOs) discharged into large lakes: Pilot-scale validation. *J. Environ. Manag.*, 256: 109937.

Brechmann, E.C. & Schepsmeier, U. 2013. Modeling dependence with C- and D-vine copulas: The R package CDVine. *J. Stat. Softw.*, 52(3): 1–27.

Brown, J.D., Spencer, T. & Moeller, I. 2007. Modeling storm surge flooding of an urban area with particular reference to modeling uncertainties: A case study of Canvey Island, United Kingdom. *Water Resour. Res.*, 43: W06402.

Bruni, G., Reinoso, R., van de Giesen, N.C., Clemens, F.H.L.R. & ten Veldhuis, J.A.E. 2015. On the sensitivity of urban hydrodynamic modelling to rainfall spatial and temporal resolution. *Hydrol. Earth Syst. Sci.*, 19(2): 691–709.

Burszta-Adamiak, E., Gajda, W., Jurczyk, A., Lewandowski, R., Licznar, P., Mańczak, P., Mikołajewski, K., Oktawiec, M., Ośródka, K., Siekanowicz, K., Stach, A., Szturc, J., Wereski, S., Wilk, P. & Zaleski, J. 2020. *Metodyka Opracowania Polskiego Atlasu Natężeń Deszczów (PANDA)*. Instytut Meteorologii i Gospodarki Wodnej Państwowy Instytut Badawczy Warszawa, Warszawa.

Butler, D. & Davies, J. (2011). *Urban Drainage*. CRC Press, London. https://doi.org/10.1201/9781315272535

Calabro, P.S. 2004. Design storms and water quality control. *J. Hydrol. Eng.*, 9(1): 28–34.

Candela, A., Brigandì, G. & Aronica, G.T. 2014. Estimation of synthetic flood design hydrographs using a distributed rainfall-runoff model coupled with a copula-based single storm rainfall generator. *Nat. Hazards Earth Syst. Sci.*, 14(7): 1819–1833.

Caradot, N., Riechel, M., Rouault, P., Caradot, A., Lengemann, N., Eckert, E., Ringe, A., Clemens, F. & Cherqui, F. 2020. The influence of condition assessment uncertainties on sewer deterioration modelling. *Struct. Infrastruct. Eng.*, 16(2): 287–296.

Caradot, N., Rouault, P., Clemens, F. & Cherqui, F. 2018. Evaluation of uncertainties in sewer condition assessment. *Struct. Infrastruct. Eng.*, 14(2): 264–273.

Chawla, I., Osuri, K.K., Mujumdar, P.P. & Niyogi, D. 2018. Assessment of the weather research and forecasting (WRF) model for simulation of extreme rainfall events in the upper Ganga Basin. *Hydrol. Earth Syst. Sci.*, 22(2): 1095–1117.

Chen, A.S., Evans, B., Djordjević, S. & Savić, D.A. 2012. Multi-layered coarse grid modelling in 2D urban flood simulations. *J. Hydrol.*, 470–471: 1–11.

Chen, Q., Dai, G. & Liu, H. 2002. Volume of Fluid model for turbulence numerical simulation of stepped spillway overflow. *J. Hydraul.*, 128(7): 683–688.

Chen, S., Garambois, P.-A., Finaud-Guyot, P., Dellinger, G., Mosé, R., Terfous, A. & Ghenaim, A. 2018. Variance based sensitivity analysis of 1D and 2D hydraulic models: An experimental urban flood case. *Environ. Model. Softw.*, 109: 167–181.

Chen, Z., Han, S., Zhou, F.Y. & Wang, K. 2013. A CFD modeling approach for municipal sewer system design optimization to minimize emissions into receiving water body. *Water Resour. Manag.*, 27: 2053–2069.

Chormański, J., Van de Voorde, T., De Roeck, T., Batelaan, O. & Canters, F. 2008. Improving distributed runoff prediction in urbanized catchments with remote sensing based estimates of impervious surface cover. *Sensors*, 8(2): 910–932.

Chow, V.T. 1959. *Open Channel Hydraulics*. McGraw-Hill Book Company, New York.

Chow, V.T. 1964. *Handbook of Applied Hydrology*. McGraw-Hill, New York.

Chow, V.T., Maidment, D.R. & Mays, L.W. 1988. *Applied Hydrology*. McGraw-Hill International Editions, New York.

Ciach, G.J. 2003. Local random errors in tipping-bucket rain gauge measurements. *J. Atmos. Ocean. Technol.*, 20(5): 752–759.

Ciach, G.J. & Krajewski, W.F. 1999. On the estimation of radar rainfall error variance. *Adv. Water Resour.*, 22(6): 585–595.

Ciepielowski, A. & Dąbkowski, Sz.L. 2006. *Metody Obliczeń Przepływów Maksymalnych w Małych Zlewniach Rzecznych (z Przykładami)*. Oficyna Wydawnicza Projprzem-EKO, Bygdoszcz.

Cinger, V.N. 1960. Transformacja maksimalnych raschdov vodochraniliščami, "Gidrome-teoizdat". Leningrad.

Clément, J.-B., Sous, D., Bouchette, F., Golay, F. & Ersoy, M. 2023. A Richards' equation-based model for wave-resolving simulation of variably-saturated beach groundwater flow dynamics. *J. Hydrol.*, 619: 129344.

Colli, M., Lanza, L.G., La Barbera, P. & Chan, P.W. 2014. Measurement accuracy of weighing and tipping-bucket rainfall intensity gauges under dynamic laboratory testing. *Atmos. Res.*, 144: 186–194.

Cosenza, A., Mannina, G., Vanrolleghem, P.A. & Neumann, M.C. 2014. Variance-based sensitivity analysis for wastewater treatment plant modeling. *Sci. Total Environ.*, 470–471: 1068–1077.

Cristiano, E., ten Veldhuis, M.-C. & van de Giesen, N. 2017. Spatial and temporal variability of rainfall and their effects on hydrological response in urban areas – a review. *Hydrol. Earth Syst. Sci.*, 21(7): 3859–3878.

Crocetti, P., Eusebi, A.L., Bruni, C., Marinelli, E., Darvini, G., Carini, C.B., Bollettini, C., Recanati, V., Akyol, Ç. & Fatone, F. 2021. Catchment-wide validated assessment of combined sewer overflows (CSOs) in a Mediterranean coastal area and possible disinfection methods to mitigate microbial contamination. *Environ Res.*, 196: 110367.

Dai, Q. & Han, D. 2014. Exploration of discrepancy between radar and gauge rainfall estimates driven by wind fields. *Water Resour. Res.*, 50(11): 8571–8588.

Dayarante, S.T. & Perera, B.J.C. 2004. Calibration of urban stormwater drainage models using hydrograph modeling. *Urban Water J.*, 1(4): 283–297.

Dąbkowski, S.L., Górska, K., Górski, J. & Szeląg, B. 2010. Wstępne wyniki badań ścieków deszczowych w jednym z kanałów w Kielcach. *Gaz Woda Techni. Sanit*, 6: 20–24.

Dąbrowski, W. 2004. *Oddziaływanie Sieci Kanalizacyjnych na Środowisko.* Wydawnictwa Politechniki Krakowskiej, Kraków.

Del Giudice, G. & Padulano, R. 2016. Sensitivity analysis and calibration of a rainfall-runoff model with the combined use of EPA-SWMM and genetic algorithm. *Acta Geophys.*, 64: 1755–1778.

De Martino, G., De Paola, F., Fontana, N., Marini, G. & Ranucci, A. 2011. Preliminary design of combined sewer overflows and stormwater tanks in Southern Italy. *Irrig. Drain.*, 60(4): 544–555.

De Paola, F. & De Martino, F. 2013. Stormwater tank performance: Design and management criteria for capture tanks using a continuous simulation and a semi-probabilistic analytical approach. *Water*, 5(4): 1699–1711.

DHI. 1998. *MOUSE – Manuál, verzia 4.0.* Danish Hydraulic Institute. Hørsholm, Dánsko.

DHI. 2000. *MOUSE Manual, Version 2000b.* Danish Hydraulic Institute, Hørsholm, Denmark.

Dirksen, J., Clemens, F.H.L.R., Korving, H., Cherqui, F., Le Gauffre, P., Ertl, T., Plihal, H., Müller, K. & Snaterse, C.T.M. 2013. The consistency of visual sewer inspection data. *Struct. Infrastruct. Eng.*, *9*(3): 214–228.

Djordjević, S., Prodanović, D., Maksimović, C., Ivetić, M. & Savić, D. 2005. SIPSON–simulation of interaction between pipe flow and surface overland flow in networks. *Water Sci. Technol.*, 52(5): 275–283.

Dong, B., Xia, J., Zhou, M., Li, Q., Ahmadian, R. & Falconer, R.A. 2022. Integrated modeling of 2D urban surface and 1D sewer hydrodynamic processes and flood risk assessment of people and vehicles. *Sci. Total Environ.*, 827: 154098.

Dotto, C.B.S., Mannina, G., Kleidorfer, M., Vezzaro, L., Henrichs, M., Mccarthy, D.T., Freni, G., Rauch, W. & Deletic, A. 2012. Comparison of different uncertainty techniques in urban stormwater quantity and quality modelling. *Water Res.*, 46(8): 2545–2558.

Duran, O., Althoefer, K. & Seneviratne, L.D. 2002. State of the art in sensor technologies for sewer inspection. *IEEE Sens. J.*, 2(2): 73–81.

DWA-A 118. 2006. *Hydraulische Bemessung und Nachweis von Entwässerungssystemen.* DWA, Hennef.

DWA-A 117. 2006. *Bemessung von Regenrückhalteräumen. Deutsche Vereinigung für Wasserwirtschaft, Abwasser und Abfall* e. V., Hennef.

Dziopak, J. 1984. *Model Matematyczny Zbiornika Retencyjnego Kanalizacji Deszczowej*, Monografia 31. Wydawnictwo Politechniki Krakowskiej, Kraków.

Dziopak, J. 1986a. Możliwości Stosowania i Obliczenie Kanalizacyjnych Zbiorników Typu CONTRACT. *Ochr. Śr.*, 488/1–2 (27–28): 45–49.

Dziopak, J. 1986b. Wielokomorowy Zbiornik Typu COMPLEMENT. *Ochr. Śr.*, 488/1–2 (27–28): 105–105.

Dziopak, J. 1988. Efektywne Sposoby Retencjonowania Ścieków w Kanalizacji. *Ochr. Śr.*, 3–4 (36–37): 35–42.

Dziopak, J. 1997. *Multi – Chamber Storage Reservoirs in the Sewerage System.* The Publishing Office of Technical University of Częstochowa, Częstochowa.

Dziopak, J. 2004. *Modelowanie Wielokomorowych Zbiorników Retencyjnych w Kanalizacji.* Oficyna Wydawnicza Politechniki Rzeszowskiej, Rzeszów.

Dziopak, J. & Słyś, D. 2006. Physical and mathematical model of gravitation – pump reservoir in sewage system. *Environ. Prot. Eng.*, 32(2): 128–137.

Dziopak, J. & Słyś D. 2007. *Modelowanie Zbiorników Klasycznych i Grawitacyjno – Pompowych w Kanalizacji*. Oficyna Wydawnicza Politechniki Rzeszowskiej, Rzeszów.

Eggert, B., Berg, P., Haerter, J.O., Jacob, D. & Moseley, C. 2015.Temporal and spatial scaling impacts on extreme precipitation. *Atmos. Chem. Phys.*, 15(10): 5957–5971.

Einfalt, T. & Wolf-Schumann, U. 1992. Beitrag zum Auffüllen lückenhafter kontinuierlicher Niederschlagsdaten. *Zeitschrift für Stadtentwässerung und Gewässerschutz*, H 19: 155–164.

Eshtawi, T., Evers, M., Tischbein, B. & Diekkrüger, B. 2016. Integrated hydrologic modeling as a key for sustainable urban water resources planning. *Water Res.*, 101: 411–428.

European Directive 2000/60/EC. 2000. Directive 2000/60/EC of the European Parliament and of the Council of 23 October 2000 establishing a framework for Community action in the field of water policy. https://eur-lex.europa.eu/legal-content/en/ALL/?uri=CELEX%3A32000L0060

Fang, T. & Ball, J.E. 2007. Evaluation of spatially variable control parameters in a complex catchment modelling system: A genetic algorithm application. *J. Hydroinform.*, 9(3): 163–173.

Fatone, F., Szeląg, B., Kiczko, A., Majerek, D., Majewska, M., Drewnowski, J. & Łagód, G. 2021. Advanced sensitivity analysis of the impact of the temporal distribution and intensity of rainfall on hydrograph parameters in urban catchments. *Hydrol. Earth Syst. Sci.*, 25(10): 5493–5516.

Fatone, F., Szeląg, B., Kowal, P., McGarity, A., Kiczko, A., Wałek, G., Wojciechowska, E., Stachura, M. & Caradot, N. 2023. An advanced tool integrating failure and sensitivity analysis into novel modeling of the stormwater flood volume. *Hydrol. Earth Syst. Sci.*, 27(18): 3329–3349.

Fletcher, T.D., Andrieu, H. & Hamel, P. 2013. Understanding, management and modelling of urban hydrology and its consequences for receiving waters: A state of the art. *Adv. Water Resour.*, 51: 261–279.

Fontanazza, C.M., Freni, G., La Loggia, G. & Notaro, V. 2011. Uncertainty evaluation of design rainfall for urban flood risk analysis. *Water Sci Technol.*, 63(11): 2641–2650.

Freni, G., Mannina, G. & Viviani, G. 2008. Uncertainty in urban stormwater quality modelling: The effect of acceptability threshold in the GLUE methodology. *Water Res.*, 42(8–9): 2061–2072.

Froehlich, D.C. 2009. Graphical sizing of small single–outlet detention basins in the Semiarid Southwest. *J. Irrig. Drain. Eng.*, 135(6): 779–790.

Fu, G. & Butler, D. 2014. Copula-based frequency analysis of overflow and flooding in urban drainage systems. *J. Hydrol.*, 510: 49–58.

Fu, G., Butler, D., Khu, S.-T. & Sun, S. 2011. Imprecise probabilistic evaluation of sewer flooding in urban drainage systems using random set theory. *Water Resour. Res.*, 47(2): W02534.

Fu, G. & Kapelan, Z. 2013. Flood analysis of urban drainage systems: Probabilistic dependence structure of rainfall characteristics and fuzzy model parameters. *J. Hydroinform.*, 15 (3): 687–699.

García-Bartual, R. & Andres-Domenech, I. 2017. A two-parameter design storm for Mediterranean convective rainfall. *Hydrol. Earth Syst. Sci.*, 21(5): 2377–2387.

Geberemariam, T. 2015. Urban drainage infrastructure design model calibration and output uncertainty minimization. *Int. J. Sci. Eng. Res.*, 3: 2347–3878.

Geiger, W. & Dreiseitl, H. 1999. *Nowe Sposoby Odprowadzania Wód Opadowych*. Oficyna Wydawnicza Projprzem-EKO, Bydgoszcz.

Ghanem, A. 2020. Optimum design of pump intake using CFD for improving hydraulic performance. *Applicat. Model Simul.*, 4: 57–63.

Graber, D.S. 2009. Generalized numerical solution for detention basin design. *J. Irrig. Drain. Eng.*, 135(4): 487–492.

Geiger, W.F., Marsalek, J. Rawls, W.J. & Zuidema, F.C. 1987. *Manual on Drainage in Urbanized Areas*, Vol.1: Planning and design of drainage systems. Reports and studies in hydrology 43, UNESCO, Paris.

Green, W.H. & Ampt, G.A. 1911. Studies on soil physics, I. The flow of air and water through soils. *J. Agric. Sci.*, 4(1): 1–24.

Grimaldi, S. & Petroselli, A. 2015. Do we still need the rational formula? An alternative empirical procedure for peak discharge estimation in small and ungauged basins. *Hydrol. Sci. J.*, 60(1): 67–77.

Grimaldi, S., Petroselli, A. & Serinaldi, F. 2012. Design hydrograph estimation in small and ungauged watersheds: Continuous simulation method versus event-based approach. *Hydrol. Process.*, 26(20): 3124–3134.

Grimaldi, S. & Serinaldi, F. 2006. Design hyetograph analysis with 3-copula function. *Hydrol. Sci. J.*, 51(2): 223–238.

Grum, M. & Aalderink, R.H. 1999. Uncertainty in return period analysis of combined sewer overflow effects using embedded Monte Carlo simulations. *Water Sci. Technol.*, 39(4): 233–240.

Guan, M., Sillanpää, N. & Koivusalo, H. 2015. Modelling and assessment of hydrological changes in a developing urban catchment. *Hydrol. Process.*, 29(13): 2880–2894.

Guo, J. & Urbonas, B. 2009. Conversion of natural watershed to kinematic wave cascading plane. *J. Hydrol. Eng.*, 14(8): 839–846.

Guo, J.C.Y. 1999. Detention basin sizing for small urban catchments. *J. Water Resour. Plan. Manag.*, 125(6): 380–382.

Guo, J.C.Y. 2000. Storm hydrographs from small urban catchments. *Water Int.*, 25(3): 481–487.

Guo, J.C.Y., Cheng, J.C.Y. & Wright, L. 2012. Field test on conversion of natural watershed into kinematic wave rectangular plane. *J. Hydrol. Eng.*, 17(8): 944–951.

Guo, K., Guan, M. & Yu, D. 2021. Urban surface water flood modelling – a comprehensive review of current models and future challenges. *Hydrol. Earth Syst. Sci.*, 25: 2843–2860.

Gupta, H.V., Sorooshian, S., & Yapo, P.O. 1999. Status of automatic calibration for hydrologic models: comparison with multilevel expert calibration. *J. Hydrol.* Eng., 4: 135–143. https://doi.org/10.1061/(ASCE)1084-0699(1999)4:2(135).

Hager, W.H. 2010. *Wastewater Hydraulics. Theory and Practice*, Second Edition.Springer Heidelberg Dordrecht London New York.

Hall, M.J. 1984. *Urban hydrology*, Elsevier Applied Science Publishers: London, New York.

Hammond, M.J., Chen, A.S., Djordjević, S., Butler, D. & Mark, O. 2015. Urban flood impact assessment: A state-of-the-art review. *Urban Water J.*, 12(1): 14–29.

Han, W.S. & Burian, S.J. 2009. Determining effective impervious area for urban hydrologic modeling. *J. Hydrol. Eng.*, 14(2): 111–120.

Hangan, H. 1999. Wind-driven rain studies. A C-FD-E approach. *J. Wind Eng. Ind. Aerodyn.*, 81(1–3): 323–331.

Hawkins, R.H. 1993. Asymptotic determination of Curve Numbers from data, *J. Irrig. Drain. Eng.*, (American Society of Civil Engineers), 119(2): 334–345.

Hegger, D.L.T., Driessen, P.P.J., Wiering, M., van Rijswick, H.F.M.W., Kundzewicz, Z.W., Matczak, P., Crabbé, A., Raadgever, G.T., Bakker, M.H.N., Priest, S.J., Larrue, C. & Ek, K. 2016. Toward more flood resilience: is a diversification of flood risk management strategies the way forward? *Ecol. Soc.*, 21(4): 52.

Hertzberg, J., Christaller, T., Kirchner, F., Licht, U. & Rome, E. Sewer robotics. In *Proceedings of the 5th International Conference Simulation Adaptive Behavior (SAB-98),* Cambridge, MA, USA, 24–26 September 1998, pp. 427–436.

Hettiarachchi, S., Wasko, C. & Sharma, A. 2018. Increase in flood risk resulting from climate change in a developed urban watershed–the role of storm temporal patterns. *Hydrol. Earth Syst. Sci.*, 22(3): 2041–2056.

Hirt, C.W. & Nichols, B.D. 1981. Volume of Fluid (VOF) method for the dynamics of free boundaries. *J. Comput. Phys.*, 39(1): 201–225.

Holman-Dodds, J.K., Bradley, A.A. & Potter, K.W. 2003. Evaluation of hydrologic benefits of infiltration based urban storm water management. *J. Am. Water Resour. Assoc.*, 39(1): 205–215.

Hong, Y.M. 2008. Graphical estimation of detention pond volume for rainfall of short duration. *J. Hydro-Environ. Res.*, 2(2): 109–117.

Horton, R.E. 1939. Analysis of runoff-plat experiments with varying infiltration capacity. *Trans. Am. Geophys. Union*, 20(4): 693–711.

Horton, R.E. 1940. An approach towards a physical interpretation of infiltration capacity. *Soil Sci. Soc. Am. J.*, Proceedings, 5(C): 399–417.

Huong, H.T.L. & Pathirana, A. 2013. Urbanization and climate change impacts on future urban flooding in Can Tho city, Vietnam. *Hydrol. Earth Syst. Sci.*, 17(1): 379–394.

Huber, W.C., Heaney, J.P., Nix, S.J., Dickinson, R. & Polmann, D. 1981. *Storm Water Managment Model. User's Manual, Version 3* (EPA-600/2-84-109a). Cincinati.

Huber, W. & Dickinson, R. 1988. *Stormwater Management Model. Version 4, Part A. User's Manual.* (EPA/600/3-88/001a) U.S. EPA, Cincinnati.

Huisman J. L. 2001. Transport and transformation process in combined sewers. *IHW Shriftenfreihe*, 10, 1–180. Zürich.

Hvitved-Jacobsen, T. 2002. SEWER PROCESSES Microbial and Chemical Process Engineering of Sewer Networks. CRC PRESS, Boca Raton, London, New York, Washington.

Ignar, S., Banasik, K. & Ignar, A. 1995. Random variability of Curve Number values for SCS runoff procedure. In *Proceedings of the International Conference on Hydrological Processes in the Catchment, held in Cracow,* Poland, 24–25 April 1995, pp. 127–130.

Iman, R.L. & Conover, W.J. 1982. A distribution-free approach to inducing rank correlation among input variables. *Commun. Stat. Simul. Comput.*, 11(3): 311–334.

IMGW–PIB. 2015. Instytut Meteorologii i Gospodarki wodnej. Państwowy Instytut Badawczy: *Instrukcja Dla Stacji Meteorologicznych.* Wydawnictwo IMGW-PIB, Warszawa.

Imhoff, K.R. & Imhoff, K.R. 1996. *Kanalizacja Miast i Oczyszczanie Ścieków. Poradnik.* Bydgoszcz Projprzem-EKO.

Iooss, B. & Lemaître, P. 2015. A review on global sensitivity analysis methods. In G. Dellino, & C. Meloni (Eds.). In *Uncertainty Management in Simulation-Optimization of Complex Systems*, pp. 101–122. Springer, New York.

Iorgulescu, I. & Beven, K.J. 2004. Nonparametric direct mapping of rainfall-runoff relationships: An alternative approach to data analysis and modelling? *Water Resour. Res.*, 40(8): 1–11.

Jacobson, C.R. 2011. Identification and quantification of the hydrological impacts of imperviousness in urban catchments: A review. *J. Environ. Manag.*, 92(6): 1438–1448.

Jato-Espino, D., Charlesworth, S.M., Bayon, J.R. & Warwick, F. 2016. Rainfall–runoff simulations to assess the potential of SuDS for mitigating flooding in highly urbanized catchments. *Int. J. Environ. Res. Publ. Health*, 13(1): 149.

Jato-Espino, D., Sillanpää, N., Andrés-Doménech, I. & Rodriguez-Hernandez, J. 2018. Flood Risk Assessment in Urban Catchments Using Multiple Regression Analysis. *J. Water Resour. Plan. Manag.*, 144(2): 04017085.

Jato-Espino, D., Sillanpää, N. & Pathak, S. 2019. Flood modelling in sewer networks using dependence measures and learning classifier systems. *J. Hydrol.*, 578: 124013.

Jiang, Y., Zevenbergena, C. & Ma, Y. 2018. Urban pluvial flooding and stormwater management: A contemporary review of China's challenges and "sponge cities" strategy. *Environ. Sci. Policy*, 80(4): 132–143.

Kaas, H., Erichsen, A.C. & Roberts, C. 2016. *Early Warning of Bathing Water Quality – An Operational Water Forecast Service*. DHI, Hørsholm, Denmark; Albany, New Zealand.

Kao, S-.C. & Govindaraju, R.S. 2007. Probabilistic structure of storm surface runoff considering the dependence between average intensity and storm duration of rainfall events. *Water Resour. Res.*, 43(6): 813–816.

Ke, Q., Tian, X. Bricker, J., Tian, Z., Guan, G., Cai, H., Huang, X., Yang, H. & Liu, J. 2020. Urban pluvial flooding prediction by machine learning approaches – a case study of Shenzen city, China. *Adv. Water Resour.*, 145: 103719.

Kiczko, A., Szeląg, B., Kozioł, A.P., Krukowski, M., Kubrak, E., Kubrak, J. & Romanowicz, R.J. 2018. Optimal capacity of a stormwater reservoir for flood peak reduction. *J. Hydrol. Eng.*, 2018, 23(4): 04018008.

Kidd, C.H.R. 1978. Rainfall Runoff Processes over Urban Catchments. *Report of Institute of Hydrology*, Wallingford, Oxon.

Kim, S.S.H., Marshall, L.A., Hughes, J.D., Sharma, A. & Vaze, J. 2021. Jointly calibrating hydrologic model parameters and state adjustments. *Water Resour. Res.*, 57(8): e2020WR028499.

Kirshen, P., Caputo, L., Vogel, R.M., Mathisen, P., Rosner, A. & Renaud, T. 2015. Adapting urban infrastructure to climate change: A drainage case study. *J. Water Resour. Plan. Manag.*, 141(4): 04014064.

Kisiel, A. 1998. *Hydrauliczna Analiza Działania Grawitacyjno-Podciśnieniowych Zbiorników Kanalizacyjnych,* Seria Inżynieria Sanitarna i Wodna, Monografia 238, Wydawnictwo Politechniki Krakowskiej, Kraków.

Kisiel, A. 2005a. Zbiornik retencyjny cieczy typu OPTIMUS – EP z podciśnieniową komorą napełnianą układem pomp ssąco – tłoczących. *Gaz Woda Tech. Sanit.*, 12: 7–10.

Kisiel, A., Kisiel, J., Malmur, R. & Mrowiec, M. 2008. Retention tanks as key elements of modern drainage systems. *Czasopismo Techniczne*, 105: 41–63.

Kisiel, A. J. 2005b. Trzykomorowe kanalizacyjne zbiorniki retencyjne typu MIRUS – EP i CONSES – EP z komorą podciśnieniową napełnianą układem pomp ssąco – tłoczących. *Gaz Woda Tech. Sanit.*, 5: 10–13.

Kleidorfer, M., Möderl, M., Sitzenfrei, R., Urich, C. & Rauch, W. 2009. A case independent approach on the impact of climate change effects on combined sewer system performance. *Water Sci. Technol.*, 60(6): 1555–1564.

Knighton, J.O. & Walter, M.T. 2016. Critical rainfall statistics for predicting watershed flood responses: Rethinking the design storm concept. *Hydrol. Process.*, 30(21): 3788–3803.

Kotowski, A. 2011. *Podstawy Bezpiecznego Wymiarowania Odwodnień Terenów*. Wydawnictwo Seidel-Przywecki, Warszawa.

Kotowski, A., Dancewicz, A. & Kazmierczak, B. 2011. Accuracy of measurements of precipitation amount using standard and tipping bucket pluviographs in comparison to Hellmann rain gauges. *Environ. Prot. Eng.*, 37(2): 23–34.

Kotowski, A., Wartalska, K. & Nowakowska, M. 2016. Uogólniona metoda analityczna wymiarowania przelewowych zbiorników retencyjnych ścieków deszczowych/Generalized analytical methods of overfall storm water retention reservoir sizing. *Ochr. Środ.*, 38: 45–52.

Krebs, G., Kokkonen, T., Valtanen, M., Koivusalo, H. & Setälä, H. 2013. A high resolution application of a stormwater management model (SWMM) using genetic parameter optimization. *Urban Water J.*, 10(6): 394–410.

Krebs, G., Kokkonen, T., Valtanen, M., Setälä, H. & Koivusalo, H. 2014. Spatial resolution considerations for urban hydrological modelling, *J. Hydrol.*, 512: 482–497.

Królikowska, J. & Królikowski A. 2012. *Wody Opadowe. Odprowadzanie, Zagospodarowanie, Podczyszczanie i Wykorzystanie.* Seidel-Przywecki, Warszawa.

Krsnik, H. 2011. *Oborinske Vode U Kanalizacijskom Sustavu Odvodnje*. Diplomski rad (Diplomma work). GFV UNIZG, Varaždin.

Kubrak, J. 1998. *Hydraulika Techniczna*. Warszawa Wydawnictwo SGGW s. 371. Warszawa.

Kusmierek-Tomaszewska, R. 2009. Porównanie wyników pomiarów opadów atmosferycznych wykonywanych standardowym deszczomierzem Hellmanna i kolebkowym stacji automatycznej /Intercomparison of precipitation measurements recorded with traditional and automated tipping-bucket rain gauges. – *Infrastruktura i Ekologia Terenów Wiejskich*, 9: 85–95.

Leandro, J., Chen, A.S., Djordjević, S. & Savić, D.A. 2009. A comparison of 1D/1D and 1D/2D coupled (sewer/surface) hydraulic models for urban flood simulation. *J. Hydraul. Eng.*, 135(6): 495–504.

Leandro, J. & Martins, R. 2016. A methodology for linking 2D overland flow models with the sewer network model SWMM 5.1 based on dynamic link libraries. *Water Sci. Technol.*, 73(3): 3017–3026.

Lee, J.G., Nietch, C.T. & Panguluri, S. 2018. Drainage area characterization for evaluating green infrastructure using the storm water management model. *Hydrol. Earth Syst. Sci.*, 22(5): 2615–2635.

Li, C. Wang, W., Xiong, J. & Chen, P. 2014a. Sensitivity analysis for urban drainage modeling using mutual information. *Entropy*, 16(11): 5738–5752.

Li, G., Sun, Y., He, Y., Li, X. & Tu, Q. 2014b. Short-term power generation energy forecasting model for small hydropower stations using GA-SVM. *Math. Probl. Eng.*, 2014(2): 1–9.

Licznar, P. 2013. Wymiarowanie zbiorników retencyjnych ścieków deszczowych na podstawie syntetycznych szeregów czasowych opadów deszczu /Reservoir dimensioning based on synthetic rainfall time series." *Ochr. Środ.*, 35(2): 27–32.

Licznar, P., Burszta-Adamiak, E., Kotowski, A., Siekanowicz-Grochowina, K. & Oktawiec, M. 2018a. Empiryczna weryfikacja modelu Bogdanowicz-Stachý do obliczania wartości natężenia deszczu miarodajnego, *Ochr. Środ.*, 40(3): 21–28.

Licznar, P., Łomotowski, J. & Kotowska, S. 2005a. Prognozowanie dobowych wartości dopływów do oczyszczalni ścieków przy wykorzystaniu sztucznych sieci neuronowych o radialnych funkcjach bazowych. *Monografie Komitetu Inżynierii Środowiska PAN*, 30: 581–592.

Licznar, P., Łomotowski, J. & Rojek, M. 2005b. *Pomiary i Przetwarzanie Danych Opadowych dla Potrzeb Projektowania i Eksploatacji Systemów Odwodnieniowych*. Futura, Poznań.

Licznar, P., Łomotowski, J. & Rupp, D.E. 2011. Random cascade driven rainfall disaggregation for urban hydrology: An evaluation of six models and a new generator. *Atmos. Res.*, 99(3–4): 563–578.

Licznar, P., Schmitt, T. & Rupp, D.E. 2011. Distributions of microcanonical cascade weights of rainfall at small timescales. *Acta Geophys.*, 59(5): 1013–1043.

Licznar, P., Siekanowicz-Grochowina, K., Oktawiec, M., Kotowski, A. & Burszta-Adamiak, E. 2018b. Empiryczna weryfikacja formuły Błaszczyka do obliczania wartości natężenia deszczu miarodajnego, *Ochr. Środ.*, 40(2): 17–22.

Linsley, R.K., Kohler, M.A. & Paulhus, J.L. 1958. *Hydrology for Engineers*. Mc Graw-Hill, New York.

Madoux-Humery, A.S., Dorner, S., Sauvé, S., Aboulfadl, K., Galarneau, M., Servais, P. & Prévost, M. 2013. Temporal variability of combined sewer overflow contaminants: Evaluation of wastewater micropollutants as tracers of fecal contamination. *Water Res.*, 47(13): 4370–4382.

Mancipe-Munoz, N., Buchberger, S.G., Suidan, M.T. & Lu, T. 2014. Calibration of rainfall-runoff model in urban watersheds for stormwater management assessment. *J. Water Resour. Plan. Manag.*, 140(6): 05014001.

Mark, O., Weesakul, S., Apirumanekul, C., Aroonnet, S.B. & Djordjević, S. 2004. Potential and limitations of 1D modelling of urban flooding. *J. Hydrol.*, 299(3–4): 284–299.

Marsalek, J., Maksimovic, C., Zeman, E. & Price, R. 1998. *Hydroinformatic Tools for Planning, Design, Operation a Rehabilitation of Sewer Systems*. NATO ASI Series - Environment, vol. 44. Springer, Dordrecht.

Marsik, M. & Waylen, P. 2006. An application of the distributed hydrologic model CASC2D to a tropical montane watershed. *J. Hydrol.*, 330(3–4): 481–495.

Martínez-Solano, F., Iglesias-Rey, P.L., Saldarriaga, J.G. & Vallejo, D. 2016. Creation of an SWMM toolkit for its application in urban drainage networks optimization. *Water*, 8(6): 259.

Mazurkiewicz, K., Skotnicki, M. & Dymaczewski, Z. 2020. Effective impervious area mapping in modeling runoff from urban catchment. *Rocznik Ochr. Środ.*, 22: 417–430.

McEnroe, B.M. 1992. Preliminary sizing of detention reservoirs to reduce peak discharges. *J. Hydraul. Eng.*, 118(11): 1540–1549.

Michelson, D., Einfalt, T., Holleman, I., Gjersten, U., Friedrich, K., Haase, G., Lindskog, M. & Jurczyk, A. 2007. *Kontrola jakości danych z radarów meteorologicznych w Europie*. Wydawnictwo Instytut Meteorologii i Gospodarki wodnej, Warszawa.

Miller, W.A. & Cunge, J.A. 1975. Simplified equations of unsteady flow. In Mahmood, K., & Yevjewich, V. (Eds.) *Unsteady Flow in Open Channels*. 367–416. Water Resources Publications, Fort Collins, CO.

Mišík, M. 1998. *Examination of the Geomorphologic and Hydrodynamic Parameters of River Hron*. (in Slovak). Water Research Institute, Bratislava.

Montserrat, A., Gutierrez, O., Poch, M. & Corominas, Ll. 2013. Field validation of a new low-cost method for determining occurrence and duration of combined sewer overflows. *Sci. Total Environ.*, 463–464: 904–912.

Moriasi, D.N., Arnold, J.G., Liew, M.W., Van Bingner, R.L., Harmel, R.D. & Veith, T.L., 2007. Model evaluation guidelines for systematic quantification of accuracy in watershed simulations. *Trans. ASABE*, 50: 885–900.

Mrowiec, M. 2002. Rurowy zbiornik przeznaczony do retencjonowania ścieków w sieci kanalizacji deszczowej. *Gaz. Woda I Tech. Sanit.*, 7: 236–239.

Mrowiec, M. 2009. *The Effective Dimensioning and Dynamic Regulation Sewage Reservoirs*. Wydawnictwo Politechniki Częstochowskiej, Częstochowa, Poland.

Muleta, M.K., McMillan, J., Amenu, G.G. & Burian, S.J. 2013. Bayesian approach for uncertainty analysis of an urban storm water model and its application to a heavily urbanized watershed. *J. Hydrol. Eng.*, 18(10): 1360–1371.

Muleta, M.K. & Nicklow, J. W. 2005. Sensitivity and uncertainty analysis coupled with automatic calibration for a distributed watershed model. *J. Hydrol.*, 306(1–4): 127–145.

Niazi, M., Nietch, C.T., Maghrebi, M., Jackson, N., Bennett, B.R., Tryby, M. & Massoudieh, A. 2017. Storm water management model: Performance review and gap analysis. *J. Sustain. Water Built Environ.*, 3(2): 04017002.

Nelder, J.A. & Mead, R. 1965. A simplex method for function minimization. *Comput. J.*, 7(4): 308–313.

Nešpor, V. & Sevruk, B. 1999. Estimation of wind-induced error of rainfall gauge measurements using a numerical simulation, *J. Atmos. Ocean. Technol.*, 16(4): 450–464.

Obermayer, A., Guenthert, F.W., Angermair, G., Tandler, R., Braunschmidt, S. & Milojevic, N. 2010. Different approaches for modelling of sewer caused urban flooding. *Water Sci. Technol.*, 62(9): 2175–2182.

Orłowski, W. 1977. *Analiza Hydraulicznej Retencji Fali Powodziowej na Zbiornikach i Dolinach Rzecznych. Informator Projektanta. Komunikat Informacyjny Centralnego Biura Studiów i Projektów Budownictwa Wodnego*. Hydroprojekt, Warszawa.

Orzechowski Z., Prywer J., Zarzycki R. 2001 Mechanika płynów w inżynierii środowiska. Wydawnictwo Naukowo-Techniczne, Warszawa.

Osorio, F., Muhaisen, O. & García, P.A. 2009. Copula-based simulation for the estimation of optimal volume for a detention basin. *J. Hydrol. Eng.*, 14(12): 1378–1382.

Paik, K. 2008. Analytical derivation of reservoir routing and hydrological risk evaluation of detention basins. *J. Hydrol.*, 352(1): 191–20.

Pang, B., Shi, S., Zhao, G., Shi, R., Peng, D. & Zhu, Z. 2020. Uncertainty assessment of urban hydrological modelling from a multiple objective perspective. *Water*, 12(5): 1393.

Papa, F., Adams, B.J. & Guo, Y. 1999. Detention time selection for stormwater quality control ponds. *Can. J. Civ. Eng.*, 26(2): 72–82.

Pecher, R. 1969. Der Abflussbeiwert Und Seine Abhängigkeit Von Der Regendauer. *Dissertation*. Fakultät für Bauwesen der Technische Hochschule München.

Pecher, R. 1972. *Das Spreicher vermögen in Misch- Und Regenwasserkanälen*. Wasser ind Boden, 12.

PN-EN 752-4: 2001. *Zewnętrzne systemy kanalizacyjne* – Obliczenia hydrauliczne i oddziaływanie na środowisko. PKN, Warszawa.

PN-EN 752:2000/2001: 2008. *Zewnętrzne systemy kanalizacyjne*. PKN, Warszawa.

PN-S-02204. 1997. *Drogi samochodowe (Odwodnienia dróg)*. PKN, Warszawa.

Pollock, M., Colli, M., O'Donnell, G., Black, A., Stagnaro, M., Dutton, M., Lanza, L., Quinn, P. & O'Connell, E. 2015. Evaluating wind-induced uncertainty on rainfall measurements by means of CFD modelling and field observations. In *Rainfall in urban and natural systems: Proceedings of the 10th International Workshop on Precipitation in Urban Areas*, Pontresina, Switzerland, pp. 181–186.

Price, R.K. & Vojinovic, Z. 2011. *Urban Hydroinformatics: Data, Models and Decision Support for Integrated Urban*. Water Management, Urban Hydroinformatics Series. IWA Publishing, London, UK.

Quijano, J.C., Zhu, Z., Morales, V., Landry, B.J. & Garcia, M.H. 2017. Three-dimensional model to capture the fate and transport of combined sewer overflow discharges: A case study in the Chicago Area Waterway System. *Sci. Total Environ.*, 576: 362–373.

Raha, D. 2007. Exploring artificial neural networks (ANN) modelling for a biological nutrient removal (BNR) sewage treatment plant (STP) to forecast effluent suspended solids. *Indian Chem. Eng.*, 49(3): 205–220.

Rezaeianzadeh, M., Tabari, H., Arabi Yazdi, A., Isik, S. & Kalin, L. 2014. Flood flow forecasting using ANN, ANFIS and regression models. *Neural Comput. Appl.*, 25(1): 25–37.

Riechel, M., Matzinger, A., Pawlowsky-Reusing, E., Sonnenberg, H., Uldack, M., Heinzmann, B., Caradot, N., von Seggern, D. & Rouault, P. 2016. Impacts of combined sewer overflows on a large urban river – understanding the effect of different management strategies. *Water Res.*, 105: 264–273.

Rodriguez-Iturbe, I., Cox, D.R. & Isham, V. 1988. A point process model for rainfall: Further developments. *Proc. R. Soc. A: Math. Phys. Eng. Sci.*, 417(1853): 283–298.

Romanowicz, R.J. & Beven, K. 2003. Estimation of flood inundation probabilities as conditioned on event inundation maps. *Water Resour. Res.*, 39(3): 4–12.

Romanowicz, R.J. & Beven, K. 2006. Comments on generalised likelihood uncertainty estimation. *Reliab. Eng. Syst. Saf.*, 91(10–11): 1315–1321.

Rossman, L.A. 2004. *Stormwater Management Model. User's Manual. Version 5.0.* National Risk Management Research Laboratory, Office of Research and Development, U. S. Environmental Research Laboratory, U. S. Environmental Protection Agency, Cincinnati, OH.

Rossman, L.A. 2009. *Storm Water Management Model. Applications Manual.* U. S. Environmental Protection Agency, Cincinnati, OH.

Rossman, L.A. 2010. Modeling low impact development alternatives with SWMM. *J. Water Manag. Model.*, 18: 167–182.

Rossman, L.A. 2015. *Storm Water Management Model: User's Manual Version 5.1.* EPA/600/R-14/413 (NTIS EPA/600/R-14/413b). U.S. Environmental Protection Agency, Cincinnati. Cincinnati, OH, USA.

Rossman, L.A. & Huber, W.C. 2016. *Storm Water Management Model Reference Manual Volume I—Hydrology* (revised) (EPA/600/R-15/162A). U.S. Environmental Protection Agency, Washington, DC, USA.

Rutkowska, A., Kohnová S., Banasik, K., Szolgay, J. & Karabová, B. 2014. Probabilistic properties of the Curve Number: A case study for small Carpathian watersheds. Manuscript submitted for consideration to be published in: *J. Mt. Sci.* 12(3): 533–548.

Rutkowski, L. 2006. *Metody i Techniki Sztucznej Inteligencji.* PWN, Warszawa.

Sadegh, M., Ragno, E. & AghaKouchak, A. 2017. Multivariate Copula Analysis Toolbox (MvCAT): Describing dependence and underlying uncertainty using a Bayesian framework. *Water Resour. Res.*, 53(6): 5166–5183.

Šalát, T. 1981. *Malá encyklopédia matematiky.* Obzor, Bratislava.

Salazar-Galán, S., García-Bartual, R., Salinas, J.L. & Francés, F. 2021. A process-based flood frequency analysis within a trivariate statistical framework. Application to a semi-arid Mediterranean case study. *J. Hydrol.* 630, Part C: 127081.

Salvadori, G. & De Michele, C. 2006. Statistical characterization of temporal structure of storms. *Adv. Water Resour.* 29(6): 827–842.

Sañudo, E., Cea, L. & Puertas, J. 2020. Modelling pluvial flooding in urban areas coupling the models Iber and SWMM. *Water*, 12(9): 2647.

Sanvicente-Sánchez, H. & Solís-Alvarado, Y. 2008. Generator of synthetic rainfall time series through markov hidden states. In Gervasi, O., Murgante, B., Laganà, A., Taniar, D., Mun, Y., & Gavrilova, M.L. (Eds.) *Computational Science and Its Applications – ICCSA 2008.* ICCSA 2008. Lecture Notes in Computer Science, vol 5073. Springer, Berlin, Heidelberg.

Schneiderman, E.M., Steenhuis, T.S., Thongs, D.J., Easton, Z.M., Zion, M.S., Neal, A.L., Mendoza, G.F. & Walter, M.T. 2007. Incorporating variable source area hydrology into a Curve-Number-based watershed model. *Hydrol. Process.*, 21(25): 3420–3430.

Sharma, D., Das Gupta, A. & Babel, M.S. 2007. Spatial disaggregation of bias-corrected GCM precipitation for improved hydrologic simulation: Ping River Basin. Thailand. *Hydrol. Earth Syst. Sci.*, 11: 1373–1390.

Shih, D-S. & Yeh, G-T. 2010. Identified model parameterization, calibration and validation of the physically distributed hydrological model, WASH123D in Taiwan. *J. Hydrol. Eng.*, 16(2). https://doi.org/10.1061/(ASCE)HE.1943-5584.0000293

SHMU. (1999–2009). *Surface Water Quality in Slovakia.* Slovak Hydrometeorological Institute, Bratislava, Slovakia. www.shmu.sk.

Shrestha, D.L., Kayastha, N. & Solomatine, D.P. 2009. A novel approach to parameter uncertainty analysis of hydrological models using neural networks. *Hydrol. Earth Syst. Sci.*, 13(7): 1235–1248.

Singh, V.P. 1996. *Kinematic Wave Modeling in Water Resources: Surface Water Hydrology.* Wiley, New York.

SNi P II-32-74. 1976. rozdz. 32. *Kanalizacja sieci zewnętrzne i budowle.* Strojizdaf, Moskva.

Skotnicki, M. & Sowiński, M. 2009. Weryfikacja metody wyznaczania szerokości hydraulicznej zlewni cząstkowych na przykładzie wybranej zlewni miejskiej. *Prace Naukowe Politechniki Warszawskiej, Inż. Śr.*, 57: 27–43.

Skotnicki, M. & Sowiński, M. 2015. Dokładność odwzorowania struktury systemu kanalizacji deszczowej na potrzeby modelowania odpływu ze zlewni miejskiej. *Gaz. Woda Tech. Sanit.*, 11: 393–397.

Słyś, D. 2009. Retencja Zbiornikowa i Sterowanie Dopływem Ścieków do Oczyszczalni. In *Monografie Komitetu Inżynierii Środowiska Polskiej Akademii Nauk.* Monografie Komitetu Inżynierii Środowiska Polskiej Akademii Nauk, Lublin.

Słyś, D. 2010. Application of numerical simulation in design of innovative Kalipso-type sewage tank. *Environ. Prot. Eng.*, 36(3): 113–126.

Sokáč, M. 2010. *The Influence of Discontinuous Pollution Sources on the Receiving Water Bodies.* Edition of scientific works of the Faculty of Civil Engineering, vol. 87. Slovak University of Technology, Bratislava.

Sokáč, M. 2019. Water balance in urban areas. *IOP Conf. Ser.: Mater. Sci. Eng.*, 471: 042028.

Song, S., Singh, V.P. 2010. Meta-elliptical copulas for drought frequency analysis of periodic hydrologic data. *Stoch Environ Res Risk Assess* (24): 425–444. https://doi.org/10.1007/s00477-009-0331-1.

Song, X., Zhang, J., Zhan, C., Xuan, Y., Ye, M. & Xu, C. 2015. Global sensitivity analysis in hydrological modeling: Review of concepts, methods, theoretical framework, and applications. *J. Hydrol.*, 523: 739–757.

Stalmann, V. 1991. Auffüllen lückenhafter Niederschlags-Datenreihen. *Zeitschrift für Stadtentwässerung und Gewässerschutz*, H 17, 135–142.

Stirrup, M. & Marchant, D. 2002. Simulation of combined sewer overflow storage tank. *J. Water Manag. Model.*, R208–R217.

Suligowski, R. 2004. Temporal and Spatial Structure of Rainfall in Poland: Regionalization Attempt. *PhD Thesis.* Warsaw University, Warsaw.

Sun, N., Hall, M., Hong, B. & Zhang, L. 2014. Impact of SWMM catchment discretization: Case study in Syracuse. New York. *J. Hydrol. Eng.*, 19(1): 223–234.

Szeląg, B. 2013. Influence of the Hydrogramme Shape on the Capacity and Selection of Drains of a Small Retention Reservoir. *PhD Thesis.* Kielce University of Technology, Kielce.

Szeląg, B. & Bąk, Ł. 2017. Probabilistic model for the annual number of storm overflow discharges in a stormwater drainage system. *Urban Water J.*, 14(6): 604–611.

Szeląg, B., Górski, J., Bąk, Ł. & Górska, K. 2013. Modelling of stormwater quantity and quality on the example of urbanized catchment in Kielce. *Ecol. Chem. Eng.*, 20(11): 1305–1316.

Szeląg, B., & Kiczko, A. 2014. The graphic method of sizing pipe reservoir for short, high-intensity rainfalls. *Ann. Warsaw Univ. Life Sci. Land Reclam.*, 46(3): 221–232.

Szeląg, B., Kiczko, A. & Dabek, L. 2016. Sensitivity and uncertainty analysis of hydrodynamic model (SWMM) for storm water runoff forecasting in an urban basin—A case study. *Ochr. Śr.*, 38(3): 15–22.

Szeląg, B., Kiczko, A. & Dąbek, L. 2019. Stormwater reservoir sizing in respect of uncertainty. *Water*, 11(2): 321.

Szeląg, B., Kiczko, A., Łagód, G. & De Paola, F. 2021a. Relationship between rainfall duration and sewer system performance measures within the context of uncertainty. *Water Resour. Manag.*, 35: 5073–5087.

Szeląg, B., Kiczko, A., Musz-Pomorska, A., Widomski, M.K., Zaburko, J., Łagód, G., Stránský, D. & Sokáč, M. 2021b. Advanced graphical–analytical method of pipe tank design integrated with sensitivity analysis for sustainable stormwater management in urbanized catchments. *Water*, 13(8): 1035.

Szeląg, B., Kiczko, A., Studziński, J. & Dąbek, L. 2018. Hydrodynamic and probabilistic modelling of storm overflow discharges. *J. Hydroinform.*, 20(5): 1100–1110.

Szeląg, B., Kiczko, A., Zaborowska, E., Mannina, G. & Mąkinia, J. 2022a. Modeling nutrient removal and energy consumption in an advanced activated sludge system under uncertainty. *J. Environ. Manage.*, 323: 116040.

Szeląg, B., Łagód, G., Musz-Pomorska, A., Widomski, M.K., Stránský, D., Sokáč, M., Pokrývková, J. & Babko, R. 2022b. Development of rainfall-runoff models for sustainable stormwater management in urbanized catchments. *Water*, 14(13): 1997.

Szeląg, B., Majerek, D., Kiczko, A., Łagód, G., Fatone, F. & McGarity, A. 2022c. Analysis of sewer network performance in context of modernization: Modeling, sensitivity, uncertainty analysis. *J. Water Resour. Plan. Manag.*, 148(12): 04022066.

Szeląg, B. & Mrowiec, M. 2016. The methods of evaluating storage volume for single-chamber reservoir in urban catchments. *Arch. Environ. Prot.*, 42(2): 20–25.

Szeląg, B., Suligowski, R., Drewnowski, J., De Paola, F., Fernandez-Morales, F.J. & Bąk, Ł. 2021c. Simulation of the number of storm overflows considering changes in precipitation dynamics and the urbanisation of the catchment area: A probabilistic approach. *J. Hydrol.*, 598: 126275.

Szeląg, B., Suligowski, R., Studziński, J. & De Paola, F. 2020. Application of logistic regression to simulate the influence of rainfall genesis on storm overflow operations: A probabilistic approach. *Hydrol. Earth Syst. Sci.*, 24: 595–614.

Szeląg, B., Zaborowska, E. & Mąkinia, J. 2023. An algorithm for selecting a machine learning method for predicting nitrous oxide emissions in municipal wastewater treatment plants. *J. Water Process Eng.*, 54: 103939.

Sztruhár, D., Sokáč, M., Holienčin, A. & Markovič, A. 2002. Comprehensive assessment of combined sewer overflows in Slovakia. *Urban Water*, 4(3): 237–243.

Szymkiewicz, R. 2000. *Modelowanie Matematyczne Przepływów w Kanałach i Rzekach.* Wydawnictwo PWN, Warszawa.

Szymkiewicz, R. 2010. *Numerical Modelling in Open Channel Hydraulics.* Springer Dordrecht Heidelberg London New York.

Tabernacki, J. 1980. *Deszczowe Zbiorniki Retencyjne w Kanalizacji, Nowa Technika w Inżynierii Sanitarnej*, seria Wodociągi i kanalizacja nr 11. Arkady, Warszawa.

Taghipour, M., Tolouei, S., Autixier, L., Prévost, M., Shakibaeinia, A. & Dorner, S. 2019. Normalized dynamic behavior of combined sewer overflow discharges for source water characterization and management. *J. Environ. Manag.*, 249: 109386.

Tarpanelli, A., Franchini, M., Brocca, L., Camici, S., Melone, F. & Moramarco, T. 2012. A simple approach for stochastic generation of spatial rainfall patterns. *J. Hydrol.*, 472–473: 63–76.

Tavakol-Davani, H., Goharian, E., Hansen, C.H., Tavakol-Davani, H., Apul, D. & Burian, S.J. 2016. How does climate change affect combined sewer overflow in a system benefiting from stormwater harvesting systems? *Sustain. Cities Soc.* 27: 430–438.

Tedela, N.H., McCutcheon, S.C., Rasmussen, T.C., Hawkins, R.H., Swank, W.T., Campbell, J.L., Adams, M.B., Jackson, C.R. & Tollner, E.W. 2012. Runoff Curve Numbers for 10 small forested watersheds in the mountains of the Eastern United States. *J. Hydrol. Eng.*, 17(11): 1188–1198.

Thorndahl, S. 2009. Stochastic long term modelling of a drainage system with estimation of return period uncertainty. *Water Sci. Technol.*, 59(12): 2331–2339.

Todeschini, S. 2016. Hydrologic and environmental impacts of imperviousness in an industrial catchment of Northern Italy. *J. Hydrol. Eng.*, 21(7): 1943–1955.

Tölgyessy, J. 1989. *Chémia, biológia a toxikológia vody a ovzdušia*, 2. Vydanie, VEDA Bratislava.

Uldack, M., Riechel, M., Heinzmann, B., Pawlowsky-Reusing, E. & Matzinger, A. 2013. *Demonstration of a Planning Instrument for Integrated and Impact Based CSO Control under Climate Change Conditions in Berlin.* Berlin, Kompetenzzentrum Wasser Berlin gGmbH.

UNESCO. 1979. *Impact of Urbanisation and Industrialisation on Water Resources Planning and Management.* UNESCO Studies and Reports in Hydrology, No. 26. Paris, UNESCO.

Urcikán, P. & Horváth, J. 1984. Synthetic design storm and its relation to Intensity-Duration-Frequency Curves. *Water Sci. Technol.*, 16(8–9): 69–83.

Van Bijnen, M., Korving, H. & Clemens, F. 2012. Impact of sewer condition on urban flooding: An uncertainty analysis based on field observations and Monte Carlo simulations on full hydrodynamic models. *Water Sci. Technol.*, 65(12): 2219–2227.

Van der Sande, C.J., de Jong, S.M. & de Roo, A.P.J. 2003. A segmentation and classification approach of IKONOS-2 imagery for land cover mapping to assist flood risk and flood damage assessment. *Int. J. Appl. Earth Obs. Geoinf.*, 4(3): 217–229.

Vandenberghe, S., Verhoest, N.E.C., Buyse, E. & De Baets, B. 2010. A stochastic design rainfall generator based on copulas and mass curves. *Hydrol. Earth Syst. Sci.*, 14(12): 2429–2442.

Vandenberghe, S., Verhoest, N.E.C., Onof, C. & De Baets, B. 2011. A comparative copula-based bivariate frequency analysis of observed and simulated storm events: A case study on Bartlett-Lewis modeled rainfall. *Water Resour. Res.*, 47(7): W07529.

Venutelli, M. 2011. Analysis of dynamic wave model for unsteady flow in an open channel. *J. Hydraul. Eng.*, 137(9): 1072–1078.

Vernieuwe, H., Vandenberghe, S., De Baets, B. & Verhoest, N.E.C. 2015. A continuous rainfall model based on vine copulas. *Hydrol. Earth Syst. Sci.*, 19(6): 2685–2699.

Verworn H.-R. 1992: Einige Kriterien für das Auffüllen lückenhafter Zeitreihen des Niederschlags. *Zeitschrift für Stadtentwässerung und Gewässerschutz*, H 18: 155–161.

Wang, J. & Guo, J.C.Y. 2018. An analytical stochastic approach for evaluating the performance of combined sewer overflow tanks. *Water Resour. Res.*, 54(5): 3357–3375.

Wang, Q., Wang, Y., Niu, R. & Peng, L. 2017. Integration of information theory, K-means cluster analysis and the logistic regression model for landslide susceptibility mapping in the Three Gorges Area, China. *Remote Sens.*, 9(9): 938.

Wang, Q.J. 1991. The genetic algorithm and its application to calibrating conceptual rainfall-runoff models. *Water Resour. Res.*, 27(9): 2467–2471.

Wang, X. & Bi, H. 2020. The effects of rainfall intensities and duration on SCS-CN model parameters under simulated rainfall. *Water*, 12(6): 1595.

Wartalska, K. & Kotowski, A. 2020. Model hyetographs of short-term rainfall for Wrocław in the perspective of 2050. *Atmosphere*, 11(6): 663.

Wilks, D.S. & Wilby, R.L. 1999. The weather generation game: A review of stochastic weather models. *Prog. Phys. Geogr.*, 23(3): 329–357.

WMO. 2011. Climate observations, stations and networks. *Guide to Climatological Practices*, WMO-No. 100, World Meteorological Organization, Geneva.

WMO. 2017. *Guidelines on the Calculation of Climate Normals*, WMO-No. 1203, World Meteorological Organization, Geneva.

Wong, G., Lambert, M., Leonard, M. & Metcalfe, A. 2010. Drought analysis using trivariate copulas conditional on climatic states. *J. Hydrol. Eng.*, 15(2): 129–141.

Wu, F.-C. & Tsang, Y.-P. 2004. Second-order Monte Carlo uncertainty/variability analysis using correlated model parameters: Application to salmonid embryo survival risk assessment. *Ecol. Model.*, 177(3–4): 393–414.

Wu, J.Y., Thompson, J.R., Kolka, R.K., Franz, K.J. & Stewart, T.W. 2013. Using the Storm Water Management Model to predict urban headwater stream hydrological response to climate and land cover change. *Hydrol. Earth Syst. Sci.*, 17(12): 4743–4758.

Yang, Q., Ma, Z. & Zhang, S. 2022. Urban pluvial flood modeling by coupling raster-based two-dimensional hydrodynamic model and SWMM. *Water*, 14(11): 1760.

Yen, B. Ch. 1986. Hydraulics of sewers. *Adv. Hydrosci.*, 14: 1–122.

Zawilski, M. 2010. Integracja zlewni zurbanizowanej w symulacji spływu ścieków deszczowych. *Gaz Gaz. Woda i Tech. Sanit.*, 6: 28–32.

Zhang, P., Liu, R., Bao, Y., Wang, J., Yu, W. & Shen, Z. 2014. Uncertainty of SWAT model at different DEM resolutions in a large mountainous watershed. *Water Res.*, 53(15): 132–144.

Zhan, X. & Huang, M.-L. 2004. ArcCN-Runoff: An ArcGIS tool for generating curve number and runoff maps. *Environ. Model. Softw.*, 19(10): 875–879.

Zhou, Y., Shen, D., Huang, N., Guo, Y., Zhang, T. & Zhang, Y. 2019. Urban flood risk assessment using storm characteristic parameters sensitive to catchment-specific drainage system. *Sci. Total Environ.*, 659: 1362–1369.

Zoppou, C. 2001. Review of urban storm water models. *Environ. Model. Softw.*, 16(3): 195–231.

Index

For Product Safety Concerns and Information please contact our EU
representative GPSR@taylorandfrancis.com
Taylor & Francis Verlag GmbH, Kaufingerstraße 24, 80331 München, Germany

www.ingramcontent.com/pod-product-compliance
Ingram Content Group UK Ltd.
Pitfield, Milton Keynes, MK11 3LW, UK
UKHW052225190926
475702UK00007B/1205
* 9 7 8 1 0 3 2 8 9 8 5 3 7 *